XGBoost for Regression Predictive Modeling and Time Series Analysis

Learn how to build, evaluate, and deploy predictive models with expert guidance

Partha Pritam Deka

Joyce Weiner

XGBoost for Regression Predictive Modeling and Time Series Analysis

Group Product Manager: Niranjan Naikwadi

Publishing Product Manager: Tejashwini R

Book Project Manager: Urvi Sharma

Senior Content Development Editor: Manikandan Kurup

Technical Editor: Sweety Pagaria

Copy Editor: Safis Editing

Proofreader: Manikandan Kurup

Indexer: Rekha Nair

Production Designer: Ponraj Dhandapani

Senior DevRel Marketing Executive: Vinishka Kalra

First published: December 2024

Production reference: 1141124

Published by Packt Publishing Ltd.

Grosvenor House

11 St Paul's Square

Birmingham

B3 1RB, UK.

ISBN 978-1-80512-305-7

www.packtpub.com

To my wife, Loni, thank you for your constant support and encouragement, and to my son, Arangan, whose joy and excitement gave me constant motivation

– Partha Pritam Deka

To Keith, thank you for the encouragement and support!

– Joyce Weiner

Foreword

In the rapidly evolving field of data science, staying ahead of the curve requires not only mastering the tools of today but also understanding the innovations shaping tomorrow. With XGBoost emerging as a possible cornerstone algorithm for predictive modeling, this book, *XGBoost for Regression Predictive Modeling and Time Series Analysis*, serves as an informative resource for both newcomers and seasoned professionals eager to harness its full potential.

XGBoost's flexibility, efficiency, and performance have made it the go-to choice for solving a wide range of real-world problems in industries such as finance, healthcare, and manufacturing. Yet, for many, XGBoost can remain a black box—powerful, but mysterious. That's where this book comes in. Partha Deka and Joyce Weiner take you on a journey, breaking down the complexities of XGBoost and making them approachable and actionable.

This book is not just a technical manual; it offers a comprehensive guide that begins with foundational concepts and builds toward advanced, real-world applications. From time series forecasting to model interpretability using cutting-edge techniques such as SHAP, LIME, and ELI5, every chapter is crafted to deepen your understanding and broaden your ability to apply XGBoost effectively.

Beginners will appreciate the clear, step-by-step explanations, while experienced data scientists will find value in the in-depth discussions on model transparency and end-to-end deployment. This dual approach ensures that readers of all levels can walk away with practical, hands-on skills, ready to implement XGBoost in real-world scenarios.

As the world increasingly leans on data to drive innovation, having the expertise to build, interpret, and deploy robust models has never been more critical. This book is a timely resource, offering a blend of theory, best practices, and coding exercises using well-known datasets such as the California Housing Price dataset. The goal is simple: empower you to move beyond basic use cases and unlock XGBoost's full potential in solving your most complex predictive challenges.

I highly recommend this book to anyone seeking to advance their knowledge of machine learning, predictive modeling, and XGBoost. It's a must-read for anyone aiming to stay ahead in this data-driven world.

Prof. Roberto V. Zicari,

Founder of ODBMS.org and Z-Inspection

Contributors

About the authors

Partha Pritam Deka is a data science leader with 15+ years of experience in semiconductor supply chain and manufacturing. As a senior staff engineer at Intel, he has led AI and machine learning teams, achieving significant cost savings and optimizations. He and his team developed a computer vision system that improved Intel's logistics, earning CSCMP Innovation Award finalist recognition. An active AI community member, Partha is a senior IEEE member and speaker at Intel's AI Everywhere conference. He also reviews for NeurIPS, contributing to AI and analytics in semiconductor manufacturing.

I would like to express my heartfelt gratitude to my wife, Loni, for her constant encouragement, love, and guidance, and Prince Shiva, for his invaluable technical review and support.

Joyce Weiner is a principal engineer with Intel Corporation. She has over 25 years of experience in the semiconductor industry, having worked in fabrication, assembly and testing, and design. Since the early 2000s, she has deployed applications that use machine learning. Joyce is a black belt in Lean Six Sigma and her area of technical expertise is the application of data science to improve efficiency. She has a BS in Physics from Rensselaer Polytechnic Institute and an MS in Optical Sciences from the University of Arizona.

My sincere thanks to Danielle Delima for connecting Partha and me. Sandra Edmonds-Thomas, Don Donato, and Pat McDonald provided encouragement and the chance to learn and excel.

About the reviewers

Ranjeeta Bhattacharya is a seasoned senior data scientist within the AI hub of the world's largest custodian bank. Her work is deeply rooted in data and involves solving complex use cases and supporting AI/ML solutions from inception to deployment. With 15+ years of experience in data science and technology consulting, Ranjeeta has held diverse roles such as software developer, solution designer, technical analyst, delivery manager, and project manager for Fortune 500 IT consulting firms globally. Her academic background includes an undergraduate degree in computer science and engineering, a master's degree in data science, and numerous certifications and publications, reflecting her dedication to lifelong learning and knowledge sharing.

Prince Chaudhary is a lead data scientist and statistician within the Technology Development group at Intel Corporation, Arizona. He plays an extremely critical role in delivering industry-leading advanced packaging by developing innovative statistical and data science methods and analytical frameworks for statistical process control systems (PCS). He is chair of Intel's assembly PCS steering committee, where he leads PCS roadmap development, reviewing PCS setup and health across all assembly manufacturing factories. He has one approved and one pending patent, has written a chapter of a book, and published research. He is an IEEE senior member and is actively involved with the IEEE Technical Committee for Emerging Technology.

Table of Contents

Part 2: Practical Applications – Data, Features, and Hyperparameters

6

Data Cleaning, Imbalanced Data, and Other Data Problems   85

7

Feature Engineering   103

8

Encoding Techniques for Categorical Features 139

9

Using XGBoost for Time Series Forecasting 175

10

Model Interpretability, Explainability, and Feature Importance with XGBoost   195

Part 3: Model Evaluation Metrics and Putting Your Model into Production

11

Metrics for Model Evaluations and Comparisons   213

12

Managing a Feature Engineering Pipeline in Training and Inference 239

13

Deploying Your XGBoost Model   249

Preface

Machine learning is an **artificial intelligence (AI)** technique that uses historical data to train a model to do either classification, putting items into groups, or prediction, estimating future values. XGBoost is a popular library for implementing machine learning with gradient-boosting algorithms. It is fast and performant, and XGBoost offers features that enable it to handle big data.

This book will give you a solid foundation for understanding machine learning and the XGBoost algorithm, and layers of practical techniques you can use when solving data science problems. We include examples that address both categorical and numeric data and classification and regression tasks and focus our attention on time-series data for the last third of the book.

Time-series data, used in forecasting for finance, supply chain management, and other industries, can pose unique challenges when training a model. With temporal data, the order of the data will impact the model results. Care must be taken to properly encode inputs to the model to handle things such as seasonal effects, or end-of-period (month, quarter, year) impacts. Although XGBoost is not designed specifically for sequential data, it can be adapted to be applied to forecasting-type problems.

Often, books and online resources only cover proof-of-concept type applications. Here, we will discuss full production deployment. We also address practical considerations such as how to monitor model performance, when to re-train a deployed model, and how to use pipelines for ease of model maintenance.

Who this book is for

This book is for data scientists and machine learning developers who want to build effective predictive models easily using XGBoost. We've set out to provide you with hands-on examples that address common challenges we've experienced when applying machine learning in our professional careers. Our goal is for this to be a practical guide and provide you with reusable code that can be applied to multiple classification and prediction tasks.

Our target audience consists of the following:

- **Data scientists**: We'll provide a deep dive into the XGBoost model, and comparisons of XGBoost to other classification and regression tree models, so you can properly select the right model for your needs. We'll discuss data cleaning and feature engineering techniques for numeric, categorical, and time-series data so that all these data types can be effectively modeled with XGBoost.

- **Machine learning developers**: We'll discuss modeling and performance metrics, which will enable you to monitor and compare models. We'll also cover XGBoost hyperparameter tuning, using XGBoost in pipelines, and model deployment practicalities.

- **Data practitioners**: This book includes hands-on walk-throughs to get you up and running on XGBoost quickly. We also provide an approachable explanation of machine learning concepts and how the XGBoost algorithm functions to give you a solid foundation.

What this book covers

Chapter 1, *An Overview of Machine Learning, Classification, and Regression*, provides a foundation for the rest of the book by introducing machine learning concepts. It covers how ensemble classification and regression tree models can be enhanced through bagging and boosting and gives an introduction to data preparation and data engineering.

Chapter 2, *XGBoost Quick Start Guide with an Iris Data Case Study*, uses the classic Iris dataset to walk through a practical example of how to use XGBoost in Python to build a classification model. At the end of the chapter, you will have code you can repurpose for similar classification problems.

Chapter 3, *Demystifying the XGBoost Paper*, provides a general overview of the XGBoost algorithm and how it works. Through examples with small datasets, you'll learn about the features and benefits of XGBoost.

Chapter 4, *Adding on to the Quick Start – Switching out the Dataset with a Housing Data Case Study*, builds on the example in *Chapter 2* and provides hands-on experience with XGBoost to make a prediction model. The intention of this chapter in combination with *Chapter 2* is to give you an understanding of what code is dataset-specific when using XGBoost and what is independent of the dataset.

Chapter 5, *Classification and Regression Trees, Ensembles, and Deep Learning Models – What's Best for Your Data?*, compares multiple algorithms and looks at performance and accuracy measurements to test and compare XGBoost to linear regression, scikit-learn gradient boosting, and random forest models. It gives a detailed explanation of the XGBoost hyperparameters and how you can change them to meet the needs of the data you are modeling.

Chapter 6, *Data Cleaning, Imbalanced Data, and Other Data Problems*, addresses common problems with real-life datasets. It covers data exploration and cleaning in depth and provides practical code examples for multiple use cases.

Chapter 7, *Feature Engineering*, explores feature engineering using a Kaggle Housing Prices dataset. You will learn common feature engineering techniques for numerical, temporal, and categorical data, applying them to the dataset.

Chapter 8, Encoding Techniques for Categorical Features, addresses the challenge of converting text data to numerical formats that can be used by machine learning models. This chapter provides practical experience with various encoding techniques.

Chapter 9, Using XGBoost for Time Series Forecasting, provides an opportunity to apply the data cleaning methods and techniques from *Chapter 6* and the feature selection from *Chapter 7* to time-series data. You'll gain practical experience in building an XGBoost model to forecast data and evaluate the prediction.

Chapter 10, Model Interpretability, Explainability, and Feature Importance with XGBoost, explores model interpretability and explainability and gives hands-on experience with extracting feature importance. It is necessary, for transparency and trust, to be able to explain how XGBoost determines its results. This chapter demonstrates five methods for model interpretation.

Chapter 11, Metrics for Model Evaluations and Comparisons, provides hands-on experience with measuring model performance and adjusting hyperparameters, building on the discussion in *Chapter 5*.

Chapter 12, Managing a Feature Engineering Pipeline in Training and Inference, expands on the concepts and code examples in *Chapters 7* and *9* to perform feature engineering for time-series data and use a pipeline to combine feature generation with model training.

Chapter 13, Deploying Your XGBoost Model, covers how to deploy your XGBoost model into a production environment. It discusses how to leverage the multithreaded and distributed compute features of XGBoost, as well as how to package your model into a container for cloud deployment. It discusses model maintenance through REST API calls, providing examples in Python.

To get the most out of this book

Basic coding knowledge and familiarity with Python, GitHub, and other DevOps tools are expected.

Software/hardware covered in the book	Operating system requirements
Python 3.8+	Windows, macOS, or Linux
Jupyter Notebook	

> **Important note**
> The versions of the technical requirements listed in each chapter of this book are what we used to generate the code. Please note that you might need to adapt the code to accommodate newer versions of the open-source libraries.

If you are using the digital version of this book, we advise you to type the code yourself or access the code from the book's GitHub repository (a link is available in the next section). Doing so will help you avoid any potential errors related to the copying and pasting of code.

Download the example code files

You can download the example code files for this book from GitHub at https://github.com/PacktPublishing/XGBoost-for-Regression-Predictive-Modeling-and-Time-Series-Analysis. If there's an update to the code, it will be updated in the GitHub repository.

We also have other code bundles from our rich catalog of books and videos available at https://github.com/PacktPublishing/. Check them out!

Conventions used

There are a number of text conventions used throughout this book.

`Code in text`: Indicates code words in text, database table names, folder names, filenames, file extensions, pathnames, dummy URLs, user input, and Twitter handles. Here is an example: "Mount the downloaded `WebStorm-10*.dmg` disk image file as another disk in your system."

A block of code is set as follows:

```
import seaborn as sns
sns.displot( irisdata, x="Species",
    discrete = True, hue="Species",
    shrink =0.8, palette="Greys" )
```

Any command-line input or output is written as follows:

```
install -c conda-forge xgboost
conda install -c anaconda pandas
```

Bold: Indicates a new term, an important word, or words that you see onscreen. For instance, words in menus or dialog boxes appear in **bold**. Here is an example: "Select **System info** from the **Administration** panel."

> **Tips or important notes**
> Appear like this.

Get in touch

Feedback from our readers is always welcome.

General feedback: If you have questions about any aspect of this book, email us at customercare@packtpub.com and mention the book title in the subject of your message.

Errata: Although we have taken every care to ensure the accuracy of our content, mistakes do happen. If you have found a mistake in this book, we would be grateful if you would report this to us. Please visit www.packtpub.com/support/errata and fill in the form.

Piracy: If you come across any illegal copies of our works in any form on the internet, we would be grateful if you would provide us with the location address or website name. Please contact us at copyright@packtpub.com with a link to the material.

If you are interested in becoming an author: If there is a topic that you have expertise in and you are interested in either writing or contributing to a book, please visit authors.packtpub.com.

Share Your Thoughts

Once you've read *XGBoost for Regression Predictive Modeling and Time Series Analysis*, we'd love to hear your thoughts! Scan the QR code below to go straight to the Amazon review page for this book and share your feedback.

https://packt.link/r/1-805-12305-X

Your review is important to us and the tech community and will help us make sure we're delivering excellent quality content.

Download a free PDF copy of this book

Thanks for purchasing this book!

Do you like to read on the go but are unable to carry your print books everywhere?

Is your eBook purchase not compatible with the device of your choice?

Don't worry, now with every Packt book you get a DRM-free PDF version of that book at no cost.

Read anywhere, any place, on any device. Search, copy, and paste code from your favorite technical books directly into your application.

The perks don't stop there, you can get exclusive access to discounts, newsletters, and great free content in your inbox daily

Follow these simple steps to get the benefits:

1. Scan the QR code or visit the link below

https://packt.link/free-ebook/9781805123057

2. Submit your proof of purchase
3. That's it! We'll send your free PDF and other benefits to your email directly

Part 1:
Introduction to Machine Learning and XGBoost with Case Studies

In this part, you will get an overview of the fundamentals of machine learning and the capabilities of XGBoost. You will achieve this through hands-on activities and an in-depth discussion of the XGBoost paper. This paper presents the capabilities of the XGBoost package for Python.

This part contains the following chapters:

- *Chapter 1, An Overview of Machine Learning, Classification, and Regression*

- *Chapter 2, XGBoost Quick Start Guide with an Iris Data Case Study*

- *Chapter 3, Demystifying the XGBoost Paper*

- *Chapter 4, Adding on to the Quick Start – Switching out the Dataset with a Housing Data Case Study*

1

An Overview of Machine Learning, Classification, and Regression

In this chapter, we will present an overview of the fundamentals of machine learning concepts. You will learn about supervised and unsupervised learning techniques, then visit classification and regression trees, and discuss ensemble models. Then you will learn about data preparation and engineering.

In this chapter we will be covering the following topics:

- Fundamentals of machine learning
- Supervised and unsupervised learning
- Classification and regression tree models
- Ensembled models – bagging vs boosting
- Data preparation and data engineering

Fundamentals of machine learning

Machine learning is a type of **artificial intelligence** (**AI**) that allows software applications to become more accurate in predicting outcomes without being explicitly programmed to do so. Machine learning algorithms use historical data as input to predict new output values. In essence, it is the science of predicting data, finding patterns in data, etc. by learning a set of algorithms from large amounts of data but not explicitly programming. There are different sets of algorithms, but machine learning algorithms are primarily of two types: supervised and unsupervised.

Supervised and unsupervised learning

In supervised learning, an algorithm learns to map the relationship between the inputs and the outputs based on a labeled dataset. A labeled dataset includes the input data (also known as features) and the corresponding output labels (also known as targets). Basically, the aim of supervised learning is to build a mapping function that can accurately predict the output for new data. Examples of supervised learning include classification and regression. Classification focuses on predicting a discrete label, while regression focusses on predicting a continuous quantity.

Unsupervised learning tries to teach an algorithm to identify patterns and structures in data without any prior knowledge of the correct labels or outputs. In unsupervised learning, the algorithm is trained to find patterns, groupings, or clusters within that data on its own. Some common examples of unsupervised learning include clustering, dimensionality reduction, and anomaly detection.

In summary, supervised learning requires labeled data with known outputs, whereas unsupervised learning requires unlabeled data without any known outputs. Supervised learning is more commonly used for prediction, classification, or regression tasks, while unsupervised learning is more commonly used for exploratory data analysis and discovering hidden patterns or insights in data.

Classification and regression decision tree models

Classification and regression trees (CART) are a type of supervised learning algorithm that can be used both for classification and regression problems.

In a classification problem, the goal is to predict the class, label, or category of a data point or an object. One example of a classification problem is to predict whether there will be customer churn or if a customer will purchase a product based on historical data.

In a regression problem, the goal is to predict a continuous numerical value, such as the price of a house based on the input features. For example, a regression CART model could be used to predict the price of a house based on input features, such as its size, location, and other relevant features.

CART models are built by recursively splitting the data into subsets based on the value of a feature that best separates the data. The algorithm chooses the feature that maximizes the separation of the classes or minimizes the variance of the target variable. The splitting process is repeated until the data are no longer able to be split further.

This process creates a tree-like structure where each internal node represents a feature or attribute, and each leaf node represents a predicted class label or a predicted continuous value. The tree can then be used to predict the class label or continuous value for new data points by following the path down the tree based on their features.

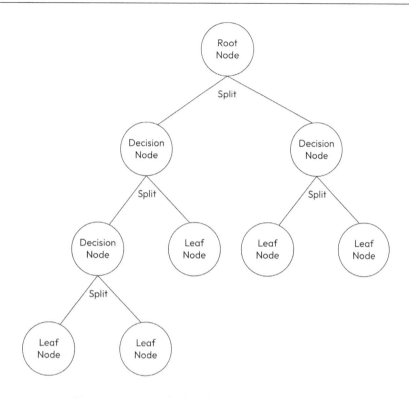

Figure 1.1 – A sample classification and regression tree

CART models are easy to explain and can handle both categorical and numerical features. However, they can be prone to overfitting. Overfitting is a phenomenon in machine learning where a model performs extremely well on the training data but fails to generalize well to unseen data. Regularization techniques such as pruning can be used to prevent overfitting. Pruning in machine learning refers to the technique of selectively removing unnecessary or less important features from a model to improve its efficiency, reduce its complexity, and prevent overfitting. The following table summarizes the advantages and disadvantages of CART models:

Advantages of CART models	Disadvantages of CART models
Easy to understand and interpret	Prone to overfitting
Relatively fast to train	Sensitive to noise in the data
Can be used for both classification and regression problems	Can be computationally expensive to train, especially for large datasets, because they need to search through all possible splits in the data in order to find the optimal tree structure

Table 1.1 – Advantages and disadvantages of CART models

As seen in the preceding table, overall, CART models are a powerful supervised learning-based tool that can be used for a variety of machine learning tasks. However, they have limitations, and we must take steps to prevent overfitting.

Ensembled models: bagging versus boosting

Ensemble modeling is a machine learning technique that combines multiple models to create a more accurate and robust model. The individual models in an ensemble are called base models. The ensemble model learns from the base models and makes predictions by combining their predictions.

Bagging and boosting are two popular ensemble learning methods used in machine learning to create more accurate models by combining individual models. However, they differ in their approach and the way they combine models.

Bagging (bootstrap aggregation) creates multiple models by repeatedly sampling the original dataset with a replacement, which means some data points may be included in multiple models, while other data points may not be included in any models. Each model is trained on its subset, and the final prediction is obtained by averaging in the case of regression or voting the predictions of all individual models in the case of classification. Since it uses a resampling technique, bagging reduces the **variance** or the impact using a different training set will have on the model.

Boosting is an iterative technique that focuses on sequentially improving the models, with each model being trained to correct the mistakes of the previous models. To begin with, a base model is trained on the entire training dataset. The subsequent models are then trained by adjusting the weights to give more importance to the misclassified instances in the previous models. The final prediction is obtained by combining the predictions of all individual models using a weighted sum, where the weights are assigned based on the performance of each model. Boosting reduces the **bias** in the model. In this context, bias means the assumptions that are being made about the form of the model function. For example, if you use a linear model, you are assuming that the form of the equation that predicts the data is linear – the model is *biased* towards linear. As you might expect, decision tree models be less biased than linear regression or logistic regression models. Boosting iterates on the equation and further reduces the bias.

The following table summarizes the key differences between bagging and boosting:

Bagging	Boosting
Models are trained individually, independently and parallelly	Models are trained sequentially, with each model trying to correct the mistakes of the previous model
Each model has equal weight in the final prediction	Each model's weight in the final prediction depends on its performance

Bagging	Boosting
Variance is reduced and overfitting removed	Bias is reduced but overfitting may occur
More accurate ensemble models are created, for example, Random Forest	More accurate ensemble models are created, for example, AdaBoost, Gradient Boosting, and XGBoost

Table 1.2 – Table summarizing the differences between bagging and boosting

The following diagram depicts the conceptual difference between bagging and boosting in a pictorial way:

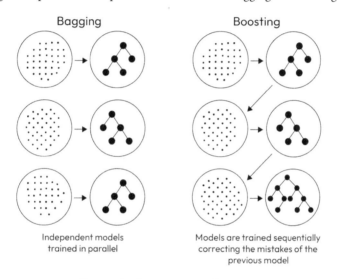

Figure 1.2 – Bagging versus boosting

Next, let's explore the two key steps in any machine learning process: data preparation and data engineering.

Data preparation and data engineering

Data preparation and data engineering are two essential steps in the machine learning process, specifically for supervised learning. We will cover each in turn in *Chapters 2* and *4*. For now, we'll provide an overview. Data preparation and data engineering involve the process of collecting, storing, and managing data so that it is accessible and useful for machine learning as well as cleaning, transforming, and formatting data so that it can be used to train and evaluate machine learning models. Lets explore and discuss some of the following topics:

1. **Collecting data**: Here, we gather data from a variety sources such as databases, sensors, or the internet.

2. **Storing data**: Here, we store data in an efficient and accessible manner. For example in SQL or NoSQL databases, file systems, etc. or others.

3. **Formatting data**: Here, we ensure that data is consistently stored in the required format. For example, data stored in tables in an SQL database, JSON format, excel format, csv format, or text format.

4. **Splitting data**: To verify your model is not overfitting, you need to test the model on part of the dataset. For this test to be effective, the model should not "know" what the testing data looks like. Data leakage is when a data cleaning step provides information about the test set to the training set, for example, if you offset all data points by the mean of all the datapoints. This is why you divide the data into a training set and a testing set using a technique called a train-test split. It should be done before moving onto to complicated data cleaning and feature engineering. The purpose of this technique is to evaluate the performance of a machine learning on unseen data. Feature engineering techniques learn parameters from the data. It is critical to learn these parameters only from the train set to avoid overfitting.

The training set is used to train the model by feeding it with input data and the corresponding output labels. The model learns patterns and relationships in the training data, which it uses to make predictions.

The testing set, however, is used to evaluate the performance of the trained model. It serves as a proxy for new, unseen data. The model makes predictions on the testing set, and the predictions are compared against the known ground truth labels. This evaluation helps assess how well the model generalizes to new data and provides an estimate of its performance.

Data cleaning

Here we identify and handle issues in the dataset that can affect the performance and reliability of machine learning models. Some of the tasks that are performed during data cleaning are:

- **Handling missing data**: Identifying and dealing with missing values by imputing them (replacing missing values with estimated values) or removing instances or features with a significant number of missing values.

- **Handling duplicate data**: Removing duplicate data from the dataset is important for the model to avoid overfitting. Duplicate values can be removed in a variety of ways, such as performing a database query to select unique rows, using Python's pandas library to drop duplicate rows, or using a statistical package such as R to remove duplicate rows. We can also handle duplicate data by keeping the duplicates but marking them as such by adding a new column with a 0 or 1 to indicate duplicates. This new column can be used by the machine learning model to avoid overfitting.

- **Handling outliers**: We must identify and address outliers, which are extreme values that deviate from the typical pattern in the data. We can either remove them or transform them to minimize the impact on the machine learning model. Domain knowledge is important in determining how best to recognize and handle outliers in the data.

- **Handling inconsistent data**: Addressing inconsistent data, such as incorrect, conflicting, or flawed values, by standardizing formats, resolving discrepancies, or using domain knowledge to correct errors.

- **Handling imbalanced data**: If there is an imbalance in the data, for example, if there are many more of one category than the others, we can use techniques such as oversampling (replicating minority class samples) or undersampling (removing majority class samples).

Feature engineering

This involves creating new features or transforming existing features into ones that are more informative and relevant to the problem to enhance the performance of machine learning algorithms. Many techniques can be used for feature engineering; it varies depending on the specifics of the dataset and the machine learning algorithms used. The following are some of the common feature engineering techniques:

- **Feature selection**: This involves selecting the most relevant features for the machine learning algorithm. There are two main types of feature selection method:

  - **Filter method**: With this method, we can select features based on their individual characteristics, such as variance or correlation with the target variable.

  - **Wrapper method**: With this method, we can select features by iteratively building and evaluating models on different subsets of features.

- **Feature extraction**: This is the process of transforming raw data into meaningful features and capturing relevant and meaningful information. The following lists some examples:

  - Extracting statistical measures, such as normalization or standardization, and other measures, such as **principal component analysis** (**PCA**), which transforms high-dimensional data into lower-dimensional space, capturing as much of the variation in the data as possible.

  - Converting categorical data into binary values, such as one-hot encoding.

  - Converting text data into numerical representations, such as bag-of-words, and text embeddings.

  - Extracting images features using techniques such as **convolution neural networks** (**CNNs**).

Let's summarize what we've covered in this chapter.

Summary

In this chapter, you were introduced to the fundamentals of machine learning, got an overview of machine learning using CART, and learned about bagging and boosting ensembled methods to improve the performance of a CART model. You were also introduced to the topics of data preparation and data engineering. The topics introduced in this chapter are the fundamentals to start machine learning, and you have just touched the tip of the iceberg. We will cover all of these topics in more depth in the following chapters.

Next, we'll go through a quick-start introduction to provide you with an example so you can apply the concepts you learned about in the next chapter.

2

XGBoost Quick Start Guide with an Iris Data Case Study

This chapter acts as a quick start guide that's designed to give you hands-on experience using XGBoost in Python. The purpose of this chapter is to get you familiar with the code so that you can train a model with XGBoost and then use that model to make a prediction (inference). By the end of this chapter, you will have built a classifier model using XGBoost and be able to use that code as a foundation for similar classification problems. In *Chapter 4*, you'll practice using XGBoost to make predictions and learn what code stays the same when switching datasets and what needs to change. In the *What's next?* section, we'll guide you to other parts of this book based on your interest in understanding more theory or practical use cases.

In this chapter, we'll cover the following main topics:

- Downloading and installing the XGBoost package
- Ingesting and exploring data
- Preparing data for modeling
- Setting up and training XGBoost
- Using XGBoost to make a prediction
- What's next?

Technical requirements

This chapter intends to be a hands-on guide, like a quick start card that comes with a new device. The code presented in this chapter is available in this book's GitHub repository: `https://github.com/PacktPublishing/XGBoost-for-Regression-Predictive-Modeling-and-Time-Series-Analysis`.

You will need the software and Python packages mentioned here to follow along with this chapter. You can install the Python packages using Anaconda. We've used `conda-forge` as the package source:

- Python 3.9 (a virtual environment is recommended):

  - XGBoost 1.7.3

  - NumPy 1.21.5

  - pandas 1.4.2

  - scikit-learn 1.2.2

  - Seaborn 0.12.2

- Anaconda

- Jupyter notebook

- VS Code

- The Iris dataset (public domain). For this chapter, you will access the dataset from scikit-learn. It is also available from `http://archive.ics.uci.edu/ml/datasets/Iris`.

Downloading and installing the XGBoost package

The first thing you need to do is prepare your Python environment. We recommend using Anaconda to install and manage your Python packages because it does a good job of handling any package-to-package dependencies. We'll also be using a virtual environment to keep all the required packages for the projects in this book contained in a single place.

> **Anaconda and conda are recommended**
>
> Anaconda checks for package dependencies during the installation process, which is important. With Python, a single package may require others to be installed and may require a specific version of those packages. For example, pandas requires NumPy. When you use Anaconda, these version checks are done at your environment level, including everything you have already installed. This avoids a problem where you might have conflicting versions. However, it can sometimes take a while for it to build a solution that addresses all the packages you've installed. This is why you should use virtual environments – you only have the needed packages for that program in the environment Anaconda is scanning. Anaconda takes longer to install packages if you have a lot of packages already installed, as you would expect. Sometimes, Anaconda takes too long to build a solution; in that case, we've been successful by doing a `pip` install for the package that was causing the delay.

We will continue to use this Anaconda virtual environment for all the examples throughout this book. Using a virtual environment helps with managing all the Python packages; only the packages you use for this code are in the environment, making it easy to transfer the code from one computer or user to another without receiving errors or missing packages. This avoids the *it works on *my* computer* problem.

Here are the steps for preparing the environment:

1. Open an Anaconda prompt and set up your environment, including installing packages. Download and install the base packages for Python and create a virtual environment called `xgboox_book_project` on your computer:

    ```
    conda create --name xgboost_book_project python==3.9
    ```

    When prompted, type `y` to proceed. Anaconda will download and install Python and create a virtual environment.

2. The next step is to activate the environment. This starts the virtual environment and separates anything we do from the main Python installation:

    ```
    conda activate xgboost_book_project
    ```

3. Next, install the packages that you will need to manipulate and model the data. This includes XGBoost, as well as pandas, which helps with handling datasets. The pandas library depends on and installs NumPy which does complex computations:

    ```
    install -c conda-forge xgboost
    conda install -c anaconda pandas
    ```

4. Install `seaborn` so that you can make graphs and visualize the data and the results of your models:

    ```
    conda install seaborn -c conda-forge
    ```

5. Now that you have set up the packages, install JupyterLab so that you can use Jupyter notebooks:

    ```
    conda install -c conda-forge jupyterlab
    ```

    Jupyter notebooks are especially helpful when you're creating models because you can easily execute code in small blocks. You can also include Markdown text in line with code to capture what you are thinking as you explore data and build models. It's like a laboratory notebook for data science. When you finalize the model that works best for the data, we recommend moving it out of a Jupyter Notebook and into a Python `.py` file. We'll discuss this more when we cover deploying models in *Chapter 13*.

With that, you have set up your Python environment with the key packages you will be using and are ready to start working with data.

Ingesting and exploring data

Before you start modeling the Iris dataset, you need to understand what it contains. To do this, you'll need to load the data into Python and use the Seaborn graphing library to visualize the data. Seaborn is easy to use and has several graph formats you can use to explore the data and see how it appears. When you do the visualization, you will also want to keep an eye out for problems in the data, such as missing values, values that don't make sense, misspellings in text data, and so on. Once you have a comfortable understanding of how the data looks, you can move on to building a model. The Iris dataset comprises the numeric measurements of iris flower characteristics, accompanied by a column specifying the corresponding iris types. You will train a model to classify the type of iris flower based on the measurements of the petals and sepals. Next, you'll get the data and bring it into a pandas DataFrame so that you can work with it.

Ingesting the Iris dataset

At this stage, you must get the Iris dataset. We are using the Iris dataset for this example because it is well-known – you may have used it before, and it is in the public domain. Another reason to use it in this quick start guide is that it's built into scikit-learn. Let's get started:

1. **Start a notebook for the model**: Start a new Jupyter Notebook file. We like to add a few notes at the start in Markdown, and we also include a header in the first code block that explains the purpose of the file, specifies my name as the author, and gives me a place to capture a revision history. Since we'll be doing math and using pandas to manage data in DataFrames, we need to include NumPy and pandas:

    ```
    # ----------------------------------------
    # filename irisclassifier.ipynb
    # purpose classify iris flowers by species based on measurements
    # of the sepal and petal
    # author Joyce Weiner
    # revision 1.0
    # revision history 1.0 - initial script
    # ----------------------------------------
    import pandas as pd
    import numpy as np
    ```

2. **Ingest the Iris dataset**: Now, load the Iris dataset from scikit-learn. It's in a module called `datasets`. We've chosen to name it `irisarray` since the data is stored as an np array:

    ```
    from sklearn import datasets
    irisarray = datasets.load_iris()
    ```

    You'll want to see what's in the dataset, so use a `print` statement to print out what's there:

    ```
    print(irisarray)
    ```

You should see the following output:

```
[6.1, 3. , 4.9, 1.8],
[6.4, 2.8, 5.6, 2.1],
[7.2, 3. , 5.8, 1.6],
[7.4, 2.8, 6.1, 1.9],
[7.9, 3.8, 6.4, 2. ],
[6.4, 2.8, 5.6, 2.2],
[6.3, 2.8, 5.1, 1.5],
[6.1, 2.6, 5.6, 1.4],
[7.7, 3. , 6.1, 2.3],
[6.3, 3.4, 5.6, 2.4],
[6.4, 3.1, 5.5, 1.8],
[6. , 3. , 4.8, 1.8],
[6.9, 3.1, 5.4, 2.1],
[6.7, 3.1, 5.6, 2.4],
[6.9, 3.1, 5.1, 2.3],
[5.8, 2.7, 5.1, 1.9],
[6.8, 3.2, 5.9, 2.3],
[6.7, 3.3, 5.7, 2.5],
[6.7, 3. , 5.2, 2.3],
[6.3, 2.5, 5. , 1.9],
[6.5, 3. , 5.2, 2. ],
[6.2, 3.4, 5.4, 2.3],
[5.9, 3. , 5.1, 1.8]]), 'target': array([0, 0, 0, 0, 0, 0, 0, 0, 0, 0, 0, 0, 0, 0, 0, 0, 0, 0, 0, 0,
0, 0, 0, 0, 0, 0, 0, 0, 0, 0, 0, 0, 0, 0, 0, 0, 0, 0, 0, 0, 0, 0,
0, 0, 0, 0, 0, 0, 1, 1, 1, 1, 1, 1, 1, 1, 1, 1, 1, 1, 1, 1, 1,
1, 1, 1, 1, 1, 1, 1, 1, 1, 1, 1, 1, 1, 1, 1, 1, 1, 1, 1, 1, 1,
1, 1, 1, 1, 1, 1, 1, 1, 1, 1, 1, 1, 2, 2, 2, 2, 2, 2, 2, 2, 2,
2, 2, 2, 2, 2, 2, 2, 2, 2, 2, 2, 2, 2, 2, 2, 2, 2,
2, 2, 2, 2, 2, 2, 2, 2, 2, 2, 2, 2, 2, 2, 2, 2, 2]), 'frame': None, 'target_names': array(['setosa', 'versicolor', 'vi
```

Figure 2.1 – Truncated output from running print(irisarray)

Notice that this dataset is comprised of multiple arrays – it's an array of arrays. The first one is data as it contains the data values. There's another array called target that contains the species of iris encoded as 0, 1, or 2. The meaning of these values is in the target_names array. The dataset includes information about the source of the data, summary statistics, references, and, toward the bottom (we had to enable a scroll bar), a very useful array called feature_names. We will use the data, target, target_names, and feature_names arrays.

Note on the results from print(irisarray)

The target_names array is at the bottom. This is a big dataset that contains lots of information. Take some time to scroll around and look at it.

---

3. **Bring the data into a pandas DataFrame and label the columns**: The data consists of multiple arrays. To build our DataFrame, combine the arrays using the np.c_ NumPy concatenate function. When you are done, you'll have a DataFrame called irisdata that contains the sepal and petal measurements, followed by a column containing the target values. As you build the DataFrame, use columns = to set the names of the columns based on the values of the feature_names array. Set the name of the target column to Species:

```
irisdata = pd.DataFrame(
    np.c_[irisarray['data'], irisarray['target']],
```

```
        columns = irisarray['feature_names'] + ['Species']
    )
```

The values in the `Species` column are integers, so change the data type of the column to match that:

```
irisdata['Species'] = irisdata['Species'].astype(int)
```

You now have a DataFrame with the sepal and petal measurements in columns, and a column for the associated species type of each row, as shown here:

```
irisdata.head()
```
[8]   ✓ 0.2s   Python

	sepal length (cm)	sepal width (cm)	petal length (cm)	petal width (cm)	Species
0	5.1	3.5	1.4	0.2	0
1	4.9	3.0	1.4	0.2	0
2	4.7	3.2	1.3	0.2	0
3	4.6	3.1	1.5	0.2	0
4	5.0	3.6	1.4	0.2	0

Figure 2.2 – First look at the irisdata DataFrame

With that, the Iris data is in a pandas DataFrame, with the measurements of the parts of the flower as columns (features), and the target value as the classification in the `Species` column. Next, we'll explore the data and get a feel for how it looks by making some graphs.

Exploring the dataset by making graphs

As mentioned previously, we like to use graphs to explore data. Since this is a well-known and well-studied dataset, you already have some information included in the np array from scikit-learn. For example, there are 150 rows of data (number of instances) and there are four attributes and a target (five columns). Let's take a closer look.

You'll want to take a look at the dataset to check for missing data or other problems. To do this, make some plots using Seaborn:

1. **Check the number of observations**: First, check that there is an equal number of observations for each of the three species types by making a bar graph using Seaborn. Make a `displot` of the data distribution for the `Species` column:

    ```
    import seaborn as sns
    sns.displot(
        irisdata, x="Species", discrete = True,
        hue="Species", shrink =0.8, palette="Greys"
    )
    ```

Here is the resulting graph:

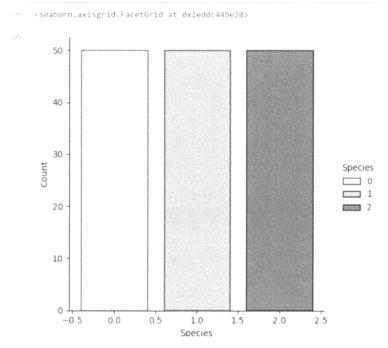

Figure 2.3 – A Seaborn displot graph that gives us insight into the number of categories in the Species column (there are three) and that there are equal numbers of examples for each category

We are setting the color palette to `Greys` so that the output on your screen matches what's printed in this book. While color is fun, it also distracts from the data. According to Edward Tufte, the leading expert in data visualization, we should strive to show the data "above all else" when making graphs (E. Tufte, *The Visual Display of Quantitative Information, Second Edition*. Cheshire, CT: Graphics Press, LLC, 2013).

In this case, the data is balanced between the three categories, with 50 observations in each group. *Chapter 6*, covers what to do if this is not the case.

2. **Plan how to visualize the numeric column data**: Now, take a look at the numeric columns. First, plan out what you want to see, then write the code to make the plots. Here, we have measurements of iris flower parts. The sepal of a flower is an outer part that encloses and protects the bud. The petals are the inner parts that attract pollinators. We are choosing to use a box and whisker plot to look at this data. A box and whisker plot provides a comparison of the data distributions between the three types of irises and allows me to see if there are any outlier

data points. Outliers are marked as diamonds on the graph. Since there are four parameters (two for sepals and two for petals), what you want is a 2x2 grid of box plots with length and width as rows and sepals and petals as columns. we like to make a quick sketch on paper before writing Python code to make our visualization:

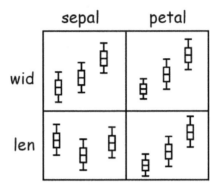

Figure 2.4 – Pencil sketch of the 2x2 box plot for the sepal and petal lengths and widths

3.   **Generate an arrangement of box plots to look at the data**: Use Matplotlib's plt.subplots package to make a grid of plots and use Seaborn to make boxplots. The first line sets up a 2x2 grid; the following calls to sns.boxplot put the plots into the grid using ax=axes[row, column]:

```
fig, axes = plt.subplots(2,2, figsize=(7,7))
sns.boxplot(ax=axes[0,0], data = irisdata,
            x="Species", y="sepal length (cm)",
            palette="Greys", hue="Species")
sns.boxplot(ax=axes[0,1], data = irisdata,
            x="Species", y="petal length (cm)",
            palette="Greys", hue="Species")
sns.boxplot(ax=axes[1,0], data = irisdata,
            x="Species", y="sepal width (cm)",
            palette="Greys", hue="Species")
sns.boxplot(ax=axes[1,1], data = irisdata,
            x="Species", y="petal width (cm)",
            palette="Greys", hue="Species")
```

Here are the output plots:

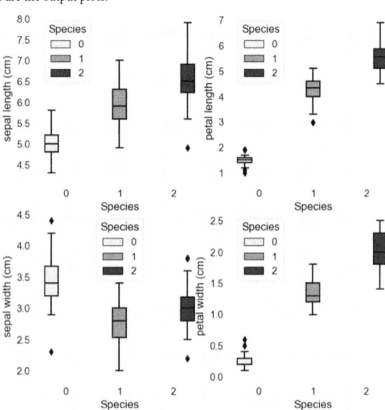

Figure 2.5 – Comparing the numeric data columns using a box plot.
Note the outlier points marked with diamonds

The data doesn't have an excessive number of outliers and appears normal and centered around the median marked by the line in the middle of the box. This means you don't have inherent challenges in this dataset. Data that is not normal or skewed requires special handling if you wish to perform statistical tests. That is not the case here.

Looking for relationships with x-y plots

Next, let's explore how the data relates to one another by making some x-y plots. This will be helpful when you model the data. You already know that in this chapter, you will be using XGBoost. Typically, however, this data exploration stage is where you begin to get ideas on what type of modeling you will want to do with the data you have. We'll talk about this more in *Chapter 5*.

To get an idea of how the data relates, use Seaborn to make an array of x-y plots comparing each of the numeric columns with each other. You will be looking for correlations between the parameters. Let's get started.

Use Seaborn to make x-y plots that compare the numeric columns: This is where Seaborn shines: we can make a single graph that compares all the numeric columns with just one line. This makes an x-y plot for each parameter versus the others. The diagonal shows the distributions. Set the background color to very light gray (efefef in hex) to make all three series visible. Then, make the graph with sns.pairplot. We've added a line to the code to make a legend:

```
sns.set_theme(
    rc={"axes.facecolor":"efefef",
    "figure.facecolor":"efefef"}
)
graphxy = sns.pairplot(irisdata,
    hue="Species",
    palette="Greys")
graphxy.add_legend()
```

Here's the graph that was produced by Seaborn showing the relationships between the parameters:

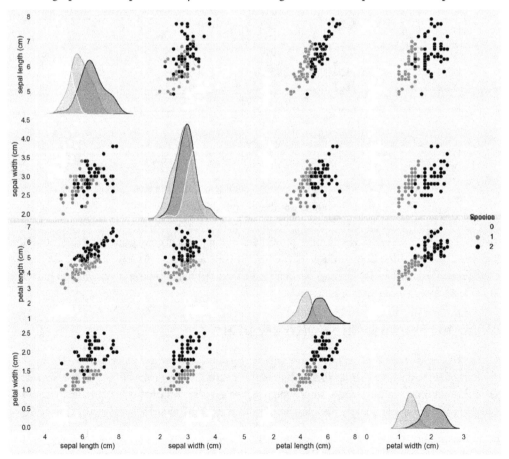

Figure 2.6 – Using a Seaborn pairplot to look for relationships in the data

In this case, we can see that there is a linear relationship between petal length and sepal length for Species 1 (iris versicolor) and Species 2 (iris virginica), but Species 0 (iris setosa) doesn't have a strong linear correlation between the two values. Rather than a line, its data is in a blob.

At this point, you have a good idea of what's in our dataset, and you've explored the data by making some quick graphs. Now, you are ready to start modeling the data.

Preparing data for modeling

To start modeling the data, you need to split the data into a training set and a testing set. You can easily do this with a function called `train_test_split` in scikit-learn. You'll want to split your data into training and testing sets at the start of the data preparation process. That way, you won't accidentally leak information from the training set into the testing set, which will inflate the results of accuracy tests. What this means is that you don't want to allow the test dataset to contain any information about the *right* answer. For example, if you calculate the mean of a dataset in full before you do the train/test split and have that data in a column in both datasets, you've provided the testing dataset with information from the training dataset. When you use your model, you want the best result, but you also don't want to have artificially increased the model's accuracy.

To use the `train_test_split` function, import it from scikit-learn. You'll want to use most of the data in the training dataset and reserve 20% for testing. To ensure you have the same rows in your datasets as in this example, use the `random_state` variable and provide the same value. The value that's picked is arbitrary. we picked 17 because we like it; there's no significance in the number. Follow these steps:

1. Split the data into a training DataFrame and a test DataFrame:

    ```
    from sklearn.model_selection import train_test_split
    training_data, testing_data = train_test_split(
        irisdata, test_size=0.2, random_state=17
    )
    ```

2. Now, we have two DataFrames – `training_data`, which contains 120 rows and 5 columns, and `testing_data`, which contains 30 rows and 5 columns. To verify the train/test split worked correctly, we can use the following code:

    ```
    training_data.shape
    (120, 5)
    ```

This checks that you have the right number of rows and columns in each split. In this example, you don't need to do further data cleaning; however, if you did, this would be the point in the process where that would happen. So far, you have prepared the dataset for modeling by splitting the data into a training set and a test set. Now, we can start building a model. You'll build the model with the training dataset, and then test the trained model using the test dataset.

> **A note on data cleaning**
>
> Data cleaning is the unappreciated part of a data science project and, at the same time, the most time-consuming. Frequently, it takes 70% to 80% of the overall project time to perform data wrangling tasks, including data cleaning. The modeling aspect, which we think of as the "project," takes less than half the overall project time. Good work during the data cleaning phase can save you lots of rework later, so it's worth taking your time to fully explore the data and clean up problem spots now.

Setting up and training XGBoost

Now that you have explored the data and checked for problems that would require data cleaning, you are ready to work with XGBoost. Next, we'll set up XGBoost to perform classification by using the measurement columns (`sepal length`, `sepal width`, `petal length`, and `petal width`) as inputs or X instances and set the species as the output or y:

1. Set the input values for the training dataset:

    ```
    X_train= training_data [[
         'sepal length (cm)','sepal width (cm)',
         'petal length (cm)','petal width (cm)'
    ]]
    X_train.head()
    ```

    This will give you the following output:

	sepal length (cm)	sepal width (cm)	petal length (cm)	petal width (cm)
29	4.7	3.2	1.6	0.2
98	5.1	2.5	3.0	1.1
37	4.9	3.6	1.4	0.1
5	5.4	3.9	1.7	0.4
81	5.5	2.4	3.7	1.0

    Figure 2.7 – Results of X_train.head()

2. Set the output (answers) to the `Species` column.

    Now, we can set up the column with the output or y value. This is the *answer* or label for the model:

    ```
    y_train = training_data[['Species']]
    y_train.head()
    ```

This will give you the following output:

	Species
29	0
98	1
37	0
5	0
81	1

Figure 2.8 – Results of y_train.head()

3. Set up the input (X) and output (y) variables for the testing data once more, separating the input columns into X_test and the output column into y_test:

```
X_test= testing_data [[
    'sepal length (cm)','sepal width (cm)',
    'petal length (cm)','petal width (cm)'
]]
y_test= testing_data [['Species']]
```

Now, you are ready to start training and check how well your model will classify what type of iris you have based on the measurements of the sepals and petals. You will train the model using the training dataset and test it with the testing set, as you'd expect. You will also do some model evaluation with built-in functions from scikit-learn to create a confusion matrix and a classification report.

4. Now, train the XGBoost classifier model using the training dataset and evaluate it with the testing dataset. First, import XGBoost and the model evaluation metrics from scikit-learn. Then, start a classifier model using the AUC evaluation and call it iris_classifier with iris_classifier = xgb.XGBClassifier(eval_metric="auc"). In the last line of this code block, you'll fit or train the model:

```
import xgboost as xgb
from sklearn.metrics import roc_auc_score
from sklearn.metrics import confusion_matrix
from sklearn.metrics import classification_report
iris_classifier = xgb.XGBClassifier(eval_metric="auc")
iris_classifier.fit(
    X_train,y_train,
    eval_set=[(X_test,y_test),(X_train,y_train)]
)
```

XGBoost evaluation metrics

XGBoost offers multiple evaluation metrics and can accept custom metrics as well. In addition to the AUC metric used here, there are over 20 options, including RMSE, mean absolute percentage error, and mean average precision. For details, see the XGBoost documentation at `https://xgboost.readthedocs.io/en/stable/parameter.html#learning-task-parameters`.

Congratulations! You now have a trained model ready to use called `iris_classifier`. This model can classify iris flowers into species based on measurements of the flower parts. You told XGBoost to use AUC, which is the receiver operating characteristic known as the area under the curve. This metric is used to decide whether to continue or stop iterating on the model. XGBoost offers multiple evaluation metrics, so you can select the best option for your specific application. A default metric is assigned based on whether the model is for prediction (the default is RMSE), classification (the default is `logloss`), or ranking (the default is map – mean average precision). You also prepared for assessing the model by importing evaluation functions.

Using XGBoost to make a prediction

At this point, you have a trained model, ready to be used to classify which type of iris you have based on the measurements of the sepals and petals on a flower. Let's test out how well it does on your test dataset. To do so, you'll need to use the `predict` method and pass it the `X_test` data:

1. Make a prediction (classify) based on the test dataset inputs and put the answers into an array called `y_score`:

```
y_score = iris_classifier.predict(X_test)
```

That's it – just one line of code to use the model! You can pass any measurement to the model, so long as you provide values for all columns: sepal length, sepal width, petal length, and petal width. Say, for example, you've measured an iris and it has a sepal length of 4.5 cm, a sepal width of 3.0 cm, a petal length of 1.5 cm, and a petal width of 0.25 cm. Which type of iris is it?

2. Next, make a prediction (classify) based on example measurements.

To use the model to do inference – in other words, make a classification from inputs – put the example data into an array called `X_example`, reshape `X_example` so that it's a two-dimensional array with 1 row and 4 columns so that it matches what your model expects for input, and then pass that array to `iris_classifier.predict`. Once you've done this, print the results of the prediction to see which type of iris we have based on the model, as shown here:

```
X_example = np.array([4.5, 3.0, 1.5, 0.25])
X_example = X_example.reshape(1,4)
y_example = iris_classifier.predict(X_example)
print(y_example)
```

The model returns a prediction of `[0]`, meaning Iris Setosa.

3. Finally, test the effectiveness of the model by calculating the accuracy, precision, recall, and F_1 score.

To finish, you should check how well the model is working. Is it putting the irises in the test dataset into the proper groups? To do this, look at the classification report. This report specifies the model's **precision**, **recall**, and **F_1 score** values, as well as the **accuracy**. This will be explained in more detail in *Chapter 11*. For now, here is a brief overview. Precision is determined by dividing the number of true positives by the sum of true positives and false positives (precision = TP/(TP+FP). Recall measures the correct classification of positives and is calculated as the number of true positives divided by the sum of true positives and false negatives (recall = TP/(TP+FN). Finally, the F_1 score is the harmonic mean of precision and recall – that is, $F_1 \ score = 2\left(\frac{precision \cdot recall}{precision + recall}\right)$:

```
print(classification_report(y_test,y_score))
```

This will print the following classification report:

	precision	recall	f1-score	support
0	1.00	1.00	1.00	7
1	0.92	1.00	0.96	11
2	1.00	0.92	0.96	12
accuracy			0.97	30
macro avg	0.97	0.97	0.97	30
weighted avg	0.97	0.97	0.97	30

Figure 2.9 – Classification report for our iris classifier

The accuracy of the model is good at 0.97, and the vast majority of the predictions are correct. The precision of 1.00 for Species 0 and Species 2 means we don't have any false positives; the model is not saying a flower is Species 0 Iris Setosa or Species 2 Iris Virginica when it isn't. The recall of 1.00 for Species 1 means there are no false negatives, meaning the model put every flower that is Iris Versicolor in the Iris Versicolor class. Finally, the F_1 score of 1.00 for Species 0 means that there are no false positives or false negatives for the Iris Setosa class. You can see this graphically by plotting a confusion matrix, something you will do as the final step of this chapter.

1. Plot the confusion matrix for the predicted values. First, we'll make a text version of the confusion matrix and print it, then we'll finish by using `ConfusionMatrixDisplay` to make a nice graphical version of the same thing. We can create the confusion matrix using the `confusion_matrix` function and pass it the true values (`y_test` – that is, the answers) and what our model predicted (`y_score`). To display the results, we can just print it:

```
conf = confusion_matrix (y_test, y_score)
print ('Confusion matrix \n', conf)
```

This produces the following output:

```
Confusion matrix
[[ 7  0  0]
 [ 0 11  0]
 [ 0  1 11]]
```

Figure 2.10 – Confusion matrix printout

2. Finally, we can make a more visually appealing version by using the `ConfusionMatrixDisplay` class and the `from_predictions` method. We'll pass it the same things – that is, the true values and what our model predicted. We've also used cmap to set the color palette:

```
ConfusionMatrixDisplay.from_predictions(
    y_test, y_score, cmap = "Greys"
)
```

Here's the output:

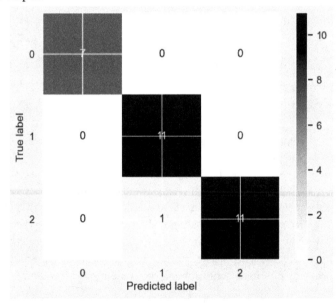

Figure 2.11 – Confusion matrix for predicting the data in the test dataset

Here, we can see that the model correctly classifies all seven examples of Species 0 Iris Setosa that are in the test dataset. It classifies all of Species 1 Iris Veritosa correctly as well. It confuses one example of Species 2 Iris Virginica for Species 1 Iris Veritosa. Overall, our model does very well.

We've used a classification model to classify iris flowers by species using the test dataset and also made a classification for example measurements. This is how you use the model once it's been deployed to do inference. Then, we verified that the model was producing reasonable results by comparing the model's output with the **ground truth** from the test dataset.

What's next?

This quick start chapter got you started with XGBoost and taught you how to build a classification model. If you're interested in building prediction models with XGBoost, in *Chapter 4*, you'll build from this quick start and create and test a predicter model. If you're interested in how XGBoost trains models and what makes XGBoost better than other gradient-boosted tree algorithms, see *Chapter 3*. *Chapter 9* covers using XGBoost on time series data.

Summary

In this quick start guide, you built a classification model using XGBoost to predict or classify irises based on their measurements. You started by downloading and installing XGBoost and other useful Python packages. Then, you ingested the data and converted it into a pandas DataFrame. You examined the data by plotting it with Seaborn to look for potential problems. Then, you split the data into a train and test set and prepared to train a model. You used the model to make predictions and classify the flowers in the test dataset. Finally, you looked at how well the model works by checking the accuracy, precision, recall, and F1 score, and by plotting a confusion matrix.

The next chapter will go into more depth on what XGBoost is doing to build that model during training and use the model during predictions.

3
Demystifying the XGBoost Paper

This chapter will simplify the XGBoost paper *XGBoost: A Scalable Tree Boosting System* to give you a general overview of the XGBoost algorithm and how it works. The paper is available on *arXiv*: `https://arxiv.org/abs/1603.02754`. We will build on what you learned in the overview of machine learning and classification and regression trees presented in *Chapter 1*.

In this chapter, you'll cover the following main topics:

- Examining the paper *XGBoost: A Scalable Tree Boosting System* at a high level
- Exploring the features and benefits of XGBoost
- Understanding the XGBoost algorithm
- Comparing XGBoost with other ensemble decision tree models

Examining the paper—XGBoost: A Scalable Tree Boosting System—at a high level

In this section, you'll review the abstract of the *XGBoost: A Scalable Tree Boosting System* paper, published in 2016, to give you a high-level overview of what to expect. This paper presents the XGBoost algorithm, which is an implementation of a gradient-boosting tree model with some tweaks. The tweaks made the model more efficient and enabled it to use more compute nodes and handle larger datasets.

The abstract starts by highlighting the demand for scalable and accurate tree-boosting methods in various fields, such as web search, ranking, and recommendations. It mentions the limitations of existing tree-boosting algorithms in terms of scalability, efficiency, and model performance.

The authors created XGBoost to address these limitations. In the abstract, they describe the key features and advantages of XGBoost, including its ability to handle large-scale data, its flexibility in defining customized optimization objectives and evaluation criteria, and its support for parallel processing and distributed computing.

The abstract mentions that XGBoost incorporates several new methods to enhance performance, such as a sparsity-aware algorithm for handling missing values and a regularization term to control model complexity. It also discusses the system's efficiency and scalability, achieved through a cache-aware block structure and parallel tree construction.

The paper includes experimental results on various datasets and demonstrates that XGBoost outperforms other tree-boosting algorithms in terms of both prediction accuracy and computational efficiency. The authors also discuss the system's impact on real-world applications and provide insights into the trade-offs between model complexity, training speed, and prediction performance.

In summary, the abstract of the paper highlights the algorithm's key features, its improvements over existing methods, and its superior performance in terms of accuracy and computational efficiency. We will examine these improvements in this chapter.

To understand what the authors have changed from previous implementations of gradient-boosted trees, and why those changes are useful, let's start with an overview of a basic gradient-boosted tree algorithm.

Exploring the features and benefits of XGBoost

In this section, you'll learn about basic gradient-boosted tree algorithms and their problems. This is so you can understand the drawbacks that exist that XGBoost was created to address. First, we will start with gradient-boosted trees and how classification and regression trees work, including the underlying algorithm of gradient descent. At the end of the section, we will summarize the problems with the previous versions of gradient-boosted tree algorithms that XGBoost fixes.

To get into the details, let's start with explaining gradient-boosted trees in simple terms by beginning with the basics of classification and regression decision trees and building up until we get to boosting.

Gradient-boosted trees

Gradient-boosted trees are a type of **classification and regression tree (CART)** model. At a high level, the model learns by building a decision tree, minimizing a loss function that compares the predicted value versus a target value. It performs this minimization by using the **gradient descent** algorithm.

Now that you have a fair idea of gradient boosting, let's get a bit more detailed in our explanation and look at how CART works.

CART

A CART model learns by building a decision tree. It can perform both classification (grouping, as you did with the iris data in *Chapter 2*) and regression (prediction).

Decision trees are built by splitting the data into subsets. At each split, the loss reduction for the left and right sides is evaluated, and the best split is selected. This means that different variables can be chosen to split the data each time, or the same variable can be chosen again. This is done again and again until splitting no longer adds value to the model.

To understand this process, let's look at a model that has been created to predict the gas mileage of a vehicle. The dataset used to create the model includes the vehicle weight, the number of cylinders in the engine, the engine displacement, and horsepower. During the training process, in *Figure 3.1*, the data has been split once based on vehicle weight, which is the best split to produce the optimum loss for both the left and right split.

The model is now calculating which parameter to use for the next split, which will partition the subset of the data with *Vehicle weight < 3400*. The algorithm selects where to split the data by looking at each parameter and calculating the amount of variation in the target parameter if the data were divided into two groups based on that parameter. In our case here, the first split was made on vehicle weight. This means that, for the full dataset, separating it into two groups, one with vehicle weight < 3400 and one with vehicle weight > 3400, made groups with the least amount of difference among the elements in those groups.

Next, the algorithm looks at the data in the *Vehicle weight < 3400* group and searches through all the parameters to find the data so that the difference in the target parameter (the variance) is the smallest. This continues until the depth of the tree reaches a threshold set in advance or until there is a minimum number of items in each group. These stopping points are set by hyperparameters for the model.

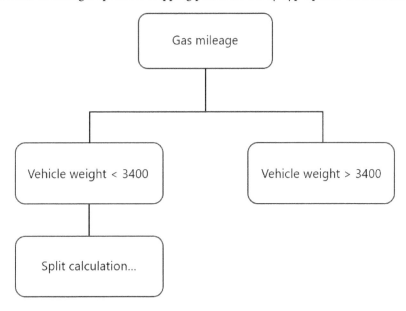

Figure 3.1 – Calculating a split in a decision tree

The algorithm works on the split calculation to further divide the left node. To evaluate the split, the algorithm minimizes the variance of the target variable.

Now you have an understanding of CART models, you're ready to layer on boosting.

Boosting

In boosting, results from multiple iterations are combined to improve the overall classification or prediction. You can think of this as repeated gradient descent algorithms. Boosting is an iterative method of combining multiple "weak learners" into a "strong learner." A weak learner is a model that is just slightly better than guessing. For a classification model with two classes, a weak learner would be a model with an accuracy of just over 50%, say 55%. When the learner is a decision tree, a model combining multiple weak learners is called a gradient-boosted tree.

To use the preceding gas mileage example, say we decide to first split on engine size and build out the tree with that decision. However, the resulting accuracy of the model is low. Using boosting, we would refine that model by repeating it, changing the parts that lead to lower accuracy. The final model uses a weighted sum of the results of the previous models based on how well they predicted the true answers.

You now have most of the pieces to understand gradient-boosted trees. To round out our explanation, we also need to simplify the process of gradient descent.

Gradient descent

Gradient descent is an algorithm that minimizes a function in an iterative manner. To picture this, imagine you are looking down into a valley with the intention of getting to the bottom. To do this, you look around from where you are currently standing and pick the steepest path because that will get you to the bottom faster. As you go lower, you again scan for the steepest path and continue in this manner until you are circling the same flat spot or reach the limit of the number of times you are willing to do the task. These limits are set at the start and are part of the algorithm hyperparameters.

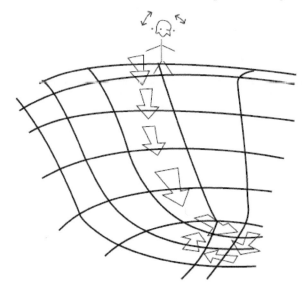

Figure 3.2 – Gradient descent – at each step, the path is scanned to find the steepest way down. This continues until you reach the bottom or run out of time

Gradient descent is used to calculate the weights for the X variables in the loss function, comparing the predicted value to the actual from the training dataset. You saw this loss function mentioned earlier. It's what we use to determine how to split the data into groups, as we discussed in the gas mileage example. We need some equation to calculate how different the members of a group are when we decide how to split the data. Gradient descent is used in regression decision trees and is part of the CART algorithm. For classification decision trees, **Gini impurity** is used in the CART algorithm. It reaches zero when all items in a node fall into a single classification.

This wraps up all the pieces and layers of an explanation of gradient-boosted trees. Next, you will learn about the existing problems that the authors of XGBoost sought to address.

> **What is Gini impurity?**
>
> Gini impurity is a way to measure how well a split has divided the data into classes. It does this by looking at the probability that a random data point from the node is classified incorrectly if randomly assigned a classification based on the distribution of classifications in the node. As the number of classes in the node reduces, the Gini impurity gets closer to zero because there is a lower probability that a random data point will be classified incorrectly. When all the members of a node are in a single class, the Gini impurity is zero.

Problems to tackle that would improve the gradient-boosted tree algorithm

Gradient-boosted tree algorithms work by iteratively splitting data as you learned above. As you can imagine, an iterative process can take a long time, especially if there is a lot of data to process. The XGBoost paper says that the XGBoost algorithm is 10 times faster than other solutions. In addition, the authors made changes that would enable scaling so XGBoost can handle large datasets.

Another difficulty with any model is potential problems with training data. These are problems you looked for in *Chapter 2*—things such as missing data, gaps and discontinuities, or overfitting that can be handled by using model fit metrics and test data. The authors added features to manage sparse data and two additional techniques to prevent overfitting. This list summarizes the key changes that were made compared to prior gradient-boosted tree algorithms. We will go into detail on each of these improvements in the next section:

- Novel tree learning algorithm
- Sparse data handling
- New algorithm for finding proposed split points
- Enabled parallel and distributed compute
- Cache-aware algorithm to prevent memory-related delays

The authors of XGBoost took steps to build in features that address problems with gradient-boosted tree algorithms. They sped up training with a new learning algorithm and added features to enable parallel and distributed computing plus reduced memory-related delay. They added features for dealing with data problems such as sparse data and overfitting. You will learn more about these new improvements in the next section.

Understanding the XGBoost algorithm

In this section, you will learn how the XGBoost algorithm tackles problems with current basic gradient-boosted tree algorithms. You will cover the improvements the authors highlight in the paper and how the improvements help correct problems. First, you will learn about how the authors addressed problems with data, then you will learn about the improvements in XGBoost that speed up training.

Addressing problems – sparse data, overfitting

To handle overfitting, a change the authors made from the standard gradient-boosted tree algorithm is to add a function (Ω, called **omega** in the paper) for the complexity of the model. This function smooths the weights to avoid overfitting. The omega function does this by penalizing complexity, meaning the algorithm prefers solutions that are simpler. This function also makes the algorithm easier to parallelize for faster computation.

Two additional techniques to handle overfitting are used:

- **Shrinkage**: In this technique, the weights are scaled after each step of the tree boosting.
- **Subsampling**: Both row subsampling and feature (column) subsampling are used. Subsampling uses a smaller set of features or rows to perform the loss calculation. This is similar to the **random forest** algorithm where the features are randomly assigned into smaller sets and a tree is built for each.

Recall that overfitting is when a model is so specific to the training data that it doesn't work well when you try to use it for other data with the same parameters.

Shrinkage helps with overfitting by reducing the weighting on some parameters, making the model more flexible. Subsampling also works to make a model more flexible by reducing the amount of data used to generate the weights for the model parameters.

One challenge for gradient-boosted tree algorithms is to handle sparse data either caused by missing data or using techniques such as one-hot encoding. One-hot encoding breaks out a column with multiple categories into multiple columns – one for each category – with the rows that belong flagged as 1 and the remaining rows flagged as 0. It creates a lot of rows with zeros in them. Datasets with lots of rows with zeros, missing or non-value data (e.g. NaN), are called sparse data. XGBoost has a sparsity-aware split-finding algorithm where there is a default split direction (left or right) set for each node. This table shows an example of sparse data:

Observation	A	B	C
1	43.2	1	0
2	33.8	0	0
3	35.1	0	1
4	38.6	0	0

...

65532	42.1	1	0
65533	40.8	0	0
65534	34.7	0	0
65535	32.3	0	0

Table 3.1– An example dataset with sparse data in columns B and C

As you see from the table, sparse data means there are a lot of rows in a column with zero or null values and only a few rows that have information. Although gradient-boosted tree algorithms work on sparse data, they are not efficient. This is because the classic gradient-boosted tree algorithm doesn't know that a row has no information and processes it as if it has information. Providing a default split direction in XGBoost speeds up the processing for sparse data by giving the algorithm a starting point to work from for the split decision. To illustrate this, let's go back to the gas mileage example. Say our dataset has sparse data because some of the values are missing:

Vehicle	Vehicle Weight	Horsepower
Sports car	3300	?
Family car	?	230
Pickup truck	6170	500

Table 3.2 – An example dataset for gas mileage prediction with sparse data

Now let's see how having a default direction works when there is sparse data. In our example, the first split we do is by vehicle weight. Additionally, the default direction is left, so we will group any missing values into the left node, and only if there is data that meets the split criteria will a data point go to the right. In our example, the sports car and family car will be grouped into the left node even though we do not have a value for the family car vehicle weight. The pickup truck will go into the right node.

For the next split on horsepower, the default direction is right, so we put the family car into the left node, and the sports car in the right node even though the sports car's horsepower is not available. The decision tree then looks like this:

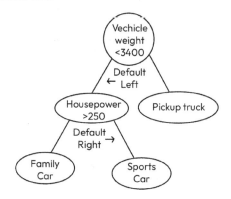

Figure 3.3 – Decision tree with default directions

So far, you have learned about enhancements in XGBoost that handle overfitting and sparse data. Next, you will learn about the enhancements the authors made to deal with memory and speed up training time.

Improvements to address memory limitations and speed up computation

XGBoost uses the exact greedy algorithm to find where to split data, which is the same algorithm that is used in **scikit-learn** and R's **gbm**. One challenge to this algorithm is that all data must fit into memory. This is a problem when using the exact greedy algorithm on very large datasets. You need very large amounts of memory to store the entire dataset at once. To address this problem, XGBoost uses an approximation. In this approximation, split candidates are first proposed based on feature distributions (what the histograms of the individual feature columns look like), then the features are mapped into buckets split by those candidates. This is illustrated in *Figure 3.4*.

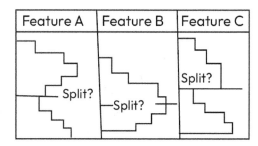

Figure 3.4 – Feature distributions

Finally, the algorithm finds the best split from the proposals based on aggregated statistics. There are two versions of the approximation – one called the global version, which uses the same split candidates from the initial mapping throughout, and one called the local version, which refines the candidates after each split.

Split candidate	Mean	Standard deviation
Feature A group 1	0.4	0.2
Feature A group 2	0.8	0.2
Feature B group 1	0.6	0.3
Feature B group 2	0.67	0.1
Feature C group 1	0.55	0.3
Feature C group 2	0.8	0.2

Table 3.3 – Illustration of split candidates and aggregated statistics for those candidates

The gradient-boosted tree algorithm requires the data to be sorted throughout the processing of the data. Sorting is a compute-intensive task. To improve the overall computational speed, the authors made modifications that optimize for the sort. Data is stored in blocks in a compressed format with each column already sorted by feature value. This helps with parallelization as these blocks can be worked on individually.

The block structure requires a fetch of gradient statistics by row index. If these statistics don't fit into the available memory cache, then there is a delay in the calculations. The approximation algorithm is made aware of the available cache size and based on that the algorithm picks the appropriate size of the data blocks, balancing parallelization and cache size. They also added features to the algorithm to reduce the amount of memory and disk access needed.

In this section, you learned that XGBoost deals with memory size limitations to allow the algorithm to handle datasets that don't all fit in memory by approximating the exact greedy algorithm to find where to split the data. You learned that the authors added a block structure to store data, which helps with parallelization, and that they improved efficiency in terms of compute by pre-sorting the blocks. They compressed the blocks and stored block-level aggregated statistics to reduce the amount of memory and disk access needed. Next, you will learn how XGBoost compares to ensemble methods other than gradient-boosted trees.

Comparing XGBoost with other ensemble decision tree models

So far in this chapter, as you've learned how XGBoost works, you've also seen how it compares to the gradient-boosted tree algorithm in scikit-learn or in R. The authors of the XGBoost paper have added features that enhance the existing algorithm and make it faster and more accurate in its predictions, or at least less prone to overfitting. Now, let's wrap up this chapter by comparing XGBoost with random forest models.

XGBoost compared with random forest models

In XGBoost, as with any gradient-boosted tree model, the decision trees are built iteratively one after another, to improve the results from the previously built trees. As a result, these decision trees are not independent. *Figure 3.5* shows how a gradient-boosted tree algorithm uses the results from the previous trees and the dataset to improve the model accuracy as measured by the training fit metrics in an iterative manner.

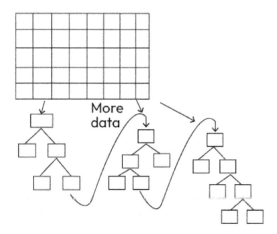

Figure 3.5 – Gradient-boosted trees are built iteratively, taking into account the results from the previously built tree

In contrast to how gradient-boosted trees are built on the previous result, the random forest algorithm builds multiple decision trees at the same time (hence the name forest). It uses a random sampling of the dataset for each tree. The results of the trees in the forest are aggregated to produce the final result. *Figure 3.6* shows how a random forest is constructed.

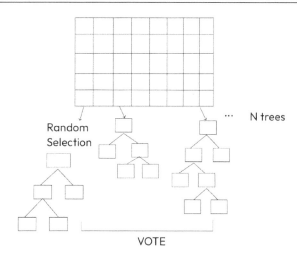

Figure 3.6 – Random Forest models build multiple trees independently,
and then aggregate the results at the end

As you can see in the preceding diagram, random forest algorithms make multiple trees, and then do the aggregation at the end to determine the final result. Gradient-boosted trees aggregate the result as they go. This means that it is more complicated to explain how the model came to its predictions or classifications with gradient-boosted trees than with random forest algorithms. Next, you will learn what to consider when deciding to use gradient-boosted trees or random forests when building a model.

When to use gradient-boosted trees

When deciding which type of decision tree model to use for your data, you should think about explainability and model performance, and also look at how the data is structured. Because gradient-boosted trees aggregate the result as they go through the training process, it is harder to explain why the model makes the prediction it does. In contrast, results from random forest models can be more easily explained because the final result is aggregated at the end, and you could look at the individual trees in the forest to explain how the model comes to its prediction. The easiest model to explain is a decision tree without any aggregation. The other factor to consider when choosing the type of decision tree model to use is how the data is structured. **Wide data** has lots of columns (features) and is well suited for random forest models where you build a number of trees on a subset of the overall features, and have them vote on the most important features. Gradient boosting works well for **tall data**, where there are a lot of rows in the dataset. Despite these data structure considerations, generally speaking, gradient boosting outperforms random forest models. Gradient-boosted tree models are, however, prone to overfitting, hence why the authors of XGBoost added features to the algorithm to address overfitting specifically. When you are building a model, take into account whether you need to explain how the model comes to its prediction and also whether the data is wide or tall. There are now explainability tools available for both random forest and XGBoost models. You will learn more about which model is best for your data in *Chapter 5*.

Summary

XGBoost is a refinement of existing gradient-boosted tree algorithms with enhancements that make it faster and less prone to overfitting. The authors of the algorithm addressed problems with sparse data by having a default direction for the splitting algorithm so rows with missing values can be handled quickly. By storing the data in sorted compressed blocks, XGBoost can scale in terms of memory and compute nodes.

In the next chapter, you'll use XGBoost to perform a prediction of house value by modifying the code you wrote in *Chapter 2*. You will change the code from a classification model to a prediction model. By the end of *Chapter 4*, you will have example code to build from for both classification and prediction models.

Adding on to the Quick Start – Switching out the Dataset with a Housing Data Case Study

This chapter builds on the work you did in *Chapter 2*. In that chapter, you built a model that *classified* irises by species based on sepal and petal measurements. In this chapter, you will build a **regression** model to predict the value of a house based on multiple parameters. You will use another famous dataset, the California housing dataset, which is from US census data in California in 1990. The intent of this example is to allow you to understand which parts of the code are dataset-specific when using XGBoost and which are the same each time. By the end of this chapter, you will have had practice using XGBoost for both classification (iris data) and regression (housing data) problems. You can then reuse the code you've written for other projects.

In this chapter, you're going to cover the following main topics:

- Switching the quick start code to a different dataset
- Preparing data for predictive modeling
- XGBoost predictive model settings and model training
- Making a prediction using XGBoost
- Comparing model parameters for the Iris and housing datasets

Technical requirements

This chapter, as with *Chapter 2*, is intended as a hands-on guide. You can use the same virtual environment already set up with the necessary packages in *Chapter 2*.

The code presented in this chapter is available on our GitHub repository: `https://github.com/PacktPublishing/XGBoost-for-Regression-Predictive-Modeling-and-Time-Series-Analysis`

You will use the following software and Python packages in this chapter:

- Python 3.9
- xgboost 1.7.3
- numpy 1.21.5
- pandas 1.4.2
- scikit-learn 1.4.2
- seaborn 0.12.2
- Anaconda
- VS Code

Switching the quick start code to a different dataset

Before you build a model, you will want to get an understanding of the housing value data and make some graphs to visualize various parameters in the dataset, just as you did with the Iris dataset. You will also want to check for problems such as missing values, values that don't make sense, and so on. In the case of the housing dataset, we have a number of parameters about the houses in a census block: average age, average number of bedrooms, average number of rooms, location, number of occupants, and a column with the median house value for that block. In the US census, a block is the smallest geographical unit for which the US Census Bureau publishes data. We will train a model to predict the value of a house given the values for the other columns.

Let's get started.

Downloading the housing dataset

Perform the following steps to load the dataset:

1. **Setting up the Python environment**: You can start by making a copy of the code from *Chapter 2* and modifying it. As in *Chapter 2*, you need Pandas and NumPy. We've chosen to call the file `housingvaluepredicter.ipynb`:

    ```
    # ----------------------------------------
    # filename housingvaluepredicter.ipynb
    # purpose predict house value based on
    # characteristics such as location, number
    # of rooms, number of bedrooms
    ```

```
# author Joyce Weiner
# revision 1.0
# revision history 1.0 - initial script
# ---------------------------------------

import pandas as pd
import numpy as np
```

2. **Ingesting the California housing data**: As with the Iris dataset, this housing dataset is built into Scikit-learn. The housing dataset is different in that the command to access it is `fetch` versus `load`. Additionally, the `fetch_california_housing` function has two useful parameters that allow you to load the data as a Pandas DataFrame and to put the feature (X) values and the target (y) values into separate variables in just one line of code:

    ```
    # load the California Housing dataset from scikit-learn
    from sklearn import datasets
    housingX, housingy = datasets.fetch_california_housing (
        return_X_y=True, as_frame=True
    )
    ```

    Now, we have two DataFrames: one called `housingX` with the feature data, and one called `housingy` with the target values. Here's what the results of `housingX.head()` look like:

	MedInc	HouseAge	AveRooms	AveBedrms	Population	AveOccup	Latitude	Longitude
0	8.3252	41.0	6.984127	1.023810	322.0	2.555556	37.88	-122.23
1	8.3014	21.0	6.238137	0.971880	2401.0	2.109842	37.86	-122.22
2	7.2574	52.0	8.288136	1.073446	496.0	2.802260	37.85	-122.24
3	5.6431	52.0	5.817352	1.073059	558.0	2.547945	37.85	-122.25
4	3.8462	52.0	6.281853	1.081081	565.0	2.181467	37.85	-122.25

Figure 4.1 – First look at the California Housing data X values we ingested in step 2

The `housingy` values look like this:

```
0    4.526
1    3.585
2    3.521
3    3.413
4    3.422
Name: MedHouseVal, dtype: float64
```

Figure 4.2 – First look at the target or y values we ingested in step 2

All about the California housing data

Before you continue, let's talk a bit about this data. Clearly, the house value data in `housingy` is not in dollars. Even back in 1990, a house in California was worth more than $4! In terms of units, the target values are given in hundreds of thousands of US dollars ($100,000). Similarly, median income is given in tens of thousands of US dollars ($10,000). The documentation for the dataset is available from the Scikit-learn website at `https://scikit-learn.org/stable/datasets/real_world.html#california-housing-dataset`.

A very important point to make about the data is that it is aggregated for a block of housing units, so the number of rooms is the total number for a block, not for an individual house. We found this page with detailed descriptions of each parameter to be helpful: `https://scikit-learn.org/dev/datasets/real_world.html#california-housing-dataset`.

We do have data for the number of households in a block. Households are defined as a group of people residing within a home unit. This means that in the case of apartment buildings, there can be many home units within a block. We should look at this more closely when we investigate the data further using graphs, which will be your next step.

You now have the housing data ingested into pandas DataFrames and are ready to visualize it with some graphs.

Exploring the dataset by making graphs

Just as you did with the Iris dataset in *Chapter 2*, you want to take a look at the dataset to check for missing data or other problems. To do this, make some plots using Seaborn. The y value is a continuous variable this time, unlike in the iris data, so you don't need to worry about whether you have the same number of examples for each class. You do want to check for discontinuity or outliers. Follow these steps:

1. **Check for discontinuity or outliers**: Use Seaborn to make a plot of the data distribution for the `housingy` data. Just as before with the iris data, We're setting the color palette to `"dark:grey"` so that the output on your screen matches what's printed in the book. This time, you want a histogram, so use the Seaborn `displot` function you used before for the iris data and set the `kind` value to `"hist"`:

```
import matplotlib.pyplot as plt
import seaborn as sns
sns.set_palette("dark:grey")
sns.displot(housingy, kind="hist")
```

This produces the following histogram:

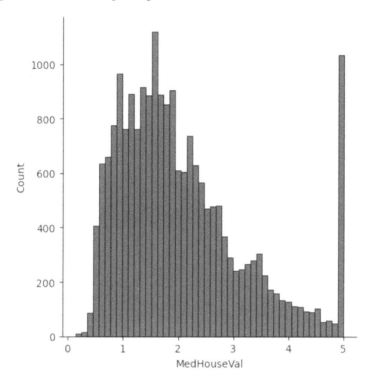

Figure 4.3 – Histogram of the median house values in housingy

This histogram shows that the dataset is bimodal, meaning there are two peaks in terms of the number of observations (counts). One peak is at approximately 1.75, which equates to a median house value of $175,000; the other is at 5, which equates to a median house value of $500,000. Notice also that there are no values above 5, which indicates the data is capped at that value.

2. **Look for relationships in the X data**: Next, look at the feature or X data. These are the input values you'll use to create a model. Since you have eight input parameters, you can make *x-y plots* and look for correlations between the input parameters. For example, you have columns for *longitude* and *latitude*, which are related. There may also be a correlation between location and house size – a neighborhood of large homes or a block of apartment buildings. To do this, use Seaborn's `pairplot`, just as you did with the iris data, and pass it `housingX`:

```
graphx = sns.pairplot(housingX)
```

This produces a matrix of graphs that compare each input parameter:

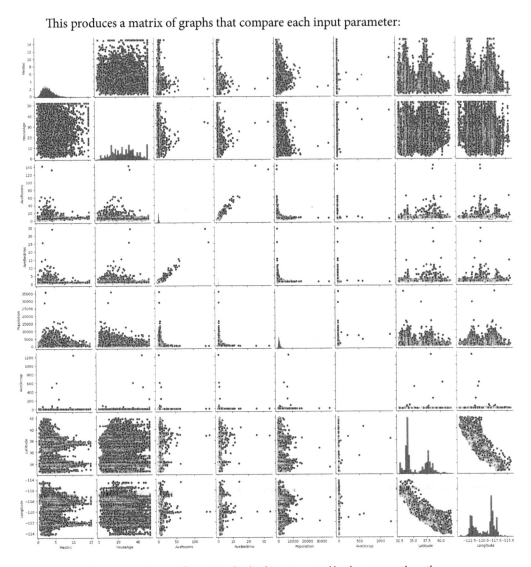

Figure 4.4 – Output from pairplot looking at every X value versus the others

Since we have 8 parameters, `pairplot` produces an 8x8 matrix. You can then see if any of the input parameters are correlated. As expected, there is a correlation between *latitude* and *longitude*. There is also a correlation between the *average number of rooms* and the *average number of bedrooms*. *Median income* appears to have some relationship with *latitude* and *longitude*, as does *population*. If input parameters are strongly correlated, you may be able to reduce the number of parameters used in our final model. We'll discuss this more in *Chapter 7*.

Other things to look for are things such as missing values, which will show as discontinuities. Look for lots of data at zero – sometimes, datasets fill missing values with zero – and check for data that is illogical or does not make sense. A thing that can happen is that errors in measurement can be indicated by values such as `9999` or `-1`; if not screened during loading to a database, these can creep into your dataset. This is why it is important to have a domain expert look over the data or have domain expertise yourself so that you can filter invalid input. In this dataset, there appears to be a cap on values for median income and house age — do you see how there is a hard line of data points at the top of each of those graphs? That is another thing to watch for.

3. **Look for correlations between the target y parameter and the input X parameters**: Just as with the iris data, you do want to look at how the input (*x*) parameters affect the target (*y*) parameter. One difference is that this time, you have a continuous variable (*median house value*) for the target, whereas the Iris dataset had a categorical parameter, the type of iris, as *y*. Another difference is that you have the data in two separate datasets rather than in one DataFrame. Having the *x* and *y* values in separate DataFrames will be helpful when you model the data with XGBoost, but to make this plot, you need to put the data together into one DataFrame. Then, you can use `pairplot` to look at the full 9x9 matrix of all parameters against each other. Concatenate the two DataFrames using pandas `pd.concat` function. In doing the concatenation, use `axis=1` to tell pandas to match up the rows using the first column, which contains the index (or row number) in each dataset. Then, make the plot with `sns.pairplot`:

```
housingxy = pd.concat([housingy, housingX], axis=1)
graphxy = sns.pairplot(housingxy)
```

This results in the following graph:

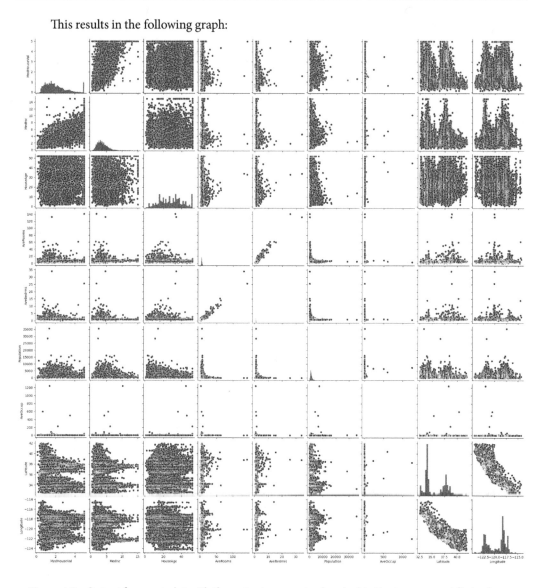

Figure 4.5 – Output from pairplot with the output parameter (y value) in the top row and first column

In the first row and column, you have the output parameter, the *median house value*. You can compare how it reacts to each of the input parameters. This lets us look at how the X values affect the output. Since we have eight X parameters, with the addition of the output, we get a 9x9 matrix. Scanning through, there is a correlation between *median income* and *median house value*, but aside from that, there aren't any striking correlations.

Domain expertise is invaluable when doing data cleaning

Knowing how the data you will be using for a model was collected is very helpful when doing data cleaning. Consider involving people who work on and with the data regularly during the data-wrangling stage of your projects. This can save a lot of time in the long run and prevent you from coming to nonsensical conclusions.

Examine your data for problems before you build models

This section has been about plotting the data. This is to look for problems before you begin modeling the data. You may find you need to collect more data points or do additional data cleaning or feature engineering based on this analysis. This is an important step, so don't skip it!

In this section, you made yourself comfortable with the data. You plotted the target values with a histogram. You made *x-y* scatter plots to look for correlations between the input variables, which would indicate you could reduce the number of inputs needed by the model. And you made an *x-y* scatter plot to look for correlations between the inputs and the target variable. Now, you can move on to building a model.

Preparing data for predictive modeling

In this section, you will prepare the data to ready it for modeling with XGBoost. To start preparing to make a model, you'll split our data into testing and training sets. As in *Chapter 2*, you will use the `train_test_split` function from Scikit-learn.

Follow these steps:

1. **Split the housing data into training and test DataFrames**: You can reuse the code from *Chapter 2*. To do this, swap `housingX` and `housingy` for `irisdata`. You can do both X and y in one line, by providing four variable names to store the results. We've chosen `X_train, X_test, y_train, y_test`. No other changes are needed from this step in *Chapter 2*. Now, you have 80% of the data in the training set, reserving 20% for testing the model. `random_state=17` is again an arbitrary value to seed the random selection of which columns go into which set; by using the same value as we do here, you'll have the same rows in your data, and our outputs will match:

```
from sklearn.model_selection import train_test_split
X_train, X_test, y_train, y_test = train_test_split(
    housingX, housingy, test_size=0.2, random_state=17
)
```

2. **Verify the split has occurred correctly**: Use `X_train.shape`, `X_test.shape`, `y_train.shape`, and `y_test.shape` to check that the number of rows and columns in each group is as expected:

    ```
    X_train.shape
    ```

    This should produce the following output:

    ```
    (16512, 8)
    ```

    The training set has 16,512 rows and 8 columns of input X values:

    ```
    X_test.shape
    ```

    This produces the following output:

    ```
    (4128, 8)
    ```

    The test set has 4128 rows (20% of the data) and 8 columns of input X values.

    ```
    y_train.shape
    ```

    This produces the following output:

    ```
    (16512,)
    ```

    The y training dataset has 16512 rows but only one column, so the number after the comma is blank:

    ```
    y_test.shape
    ```

    This produces the following output:

    ```
    (4128,)
    ```

    The y test dataset has 4128 rows and, as with the training data, only one column, so the number after the comma is blank.

3. As a final verification, use `X_test.head()` to look at the `X_test` DataFrame contents:

    ```
    X_test.head()
    ```

    The result of this statement is the following:

	MedInc	HouseAge	AveRooms	AveBedrms	Population	AveOccup	Latitude	Longitude
18403	5.4741	24.0	6.587799	1.060181	4017.0	3.311624	37.24	-121.84
13405	1.9583	7.0	5.362805	1.442073	1064.0	3.243902	34.10	-117.46
9539	2.6058	40.0	3.885714	0.914286	785.0	4.485714	37.39	-120.71
8668	4.5817	30.0	4.146135	1.108696	1526.0	1.842995	33.82	-118.39
9019	8.0137	9.0	7.734673	1.041211	12873.0	3.274739	34.16	-118.78

Figure 4.6 – Results of X_test.head()

You should see randomly ordered indices in the very first column, followed by the data for those rows.

You have split the data into test and training datasets, and in this case, your data is ready for modeling. If you need to do any data cleaning, this is when you should do it. Remember to perform the same cleaning steps on both the test and training datasets. A data pipeline is helpful because you can build it once and then apply it to test and training datasets.

Now, you are ready to build a model for this dataset.

XGBoost predictive model settings and model training

Now, you are ready to do some training and check how well your model will predict housing value based on the eight input parameters. Just as in *Chapter 2*, you'll train the model using the training dataset and test with the testing set, as you'd expect. In this section, you'll do some model evaluation with built-in functions from Scikit-learn. This time, you're performing regression rather than classification, so you won't use a confusion matrix; instead, you will look at how well your model fits the data with metrics such as R^2 and **root mean square error** (**RMSE**).

Let's talk about these metrics for a second. When you fit a model to data, you want to be sure that your model accurately represents the training data and at the same time is flexible to provide accurate predictions for data that is not in the training dataset. R^2 and RMSE measure how far the modeled values are from the actual values.

R^2 measures how much of the variance of the target is predicted by the model. An R^2 of 1 means that the predicted values match the actual values exactly. An R^2 of 0 means there is no linear relationship between the predicted and actual values. An R^2 of 0.5 means that the model can account for half of the variance in the target variable. Sometimes R^2 is written as a percentage; for example, 50% for 0.5. To calculate R^2 you divide the sum of squares of the residuals, the explained variation, by the total sum of squares, which is the total variation. The equation for R^2 is the following:

$$R^2 = 1 - \frac{\sum_i (y_i - \hat{y}_i)^2}{\sum_i (y_i - \bar{y}_i)^2}$$

Here, i = index variable, y_i = actual value, $\hat{y}_i$ = predicted value, and $\bar{y}_i$ is the mean of the actuals:

$$\bar{y} = \frac{1}{n} \sum_{i=1}^{n} y_i$$

Now that you understand how R^2 is calculated, let's discuss RMSE.

Unlike R^2, for RMSE, a smaller value is better. This is because RMSE measures the amount of error or difference between the model and the actual values. The equation for RMSE is the following:

$$RMSE = \sqrt{\left(\frac{\sum_{i=1}^{N} (y_i - \hat{y}_i)^2}{N} \right)}$$

Here, i = index variable, N = number of non-missing data points, y_i = actual value, and $\hat{y}_i$ = predicted value. This is the standard deviation of the residuals. As with R^2, RMSE measures the goodness of fit. Unlike R^2, it is not on a standardized scale and can range from 0 to infinity.

Now that you have an understanding of how to measure the performance of a regression model, we're ready to train an XGBoost model on the housing dataset and check how well it predicts housing value. Follow these steps:

1. **Train the XGBoost regression model**: Use the training dataset and evaluate the model with the testing dataset. First, import xgboost, r2_score, and root_mean_squared_error evaluation metrics from Scikit-learn. Use the default settings for xgb.XGBRegressor() prediction, which uses gbtree as a booster. This means your model will use gradient-boosted trees to define the relationship between the X values and the y output. We've chosen to call the model housevalue_regressor. By default, the XGBoost regressor uses **RMSE** to evaluate the model as it is trained:

```
import xgboost as xgb
from sklearn.metrics import r2_score
from sklearn.metrics import root_mean_squared_error
housevalue_regressor = xgb.XGBRegressor()
```

2. Then, fit (train) your model by applying the fit method, and pass the [(X_test, y_test), (X_train, y_train)] list as the evaluation set:

```
housevalue_regressor.fit(
    X_train,y_train,
    eval_set=[(X_test, y_test),(X_train,y_train)]
)
```

You now have a trained regression model called housevalue_regressor that will predict housing value based on the eight parameters in the input columns. Next, try using the model to make predictions (inference), and test how well it performs.

Making a prediction using XGBoost

You have your trained regression model and now need to use it to predict the output values based on the inputs in our test dataset. Follow these steps:

1. Use the model to make predictions of house value based on the test dataset and put the answers into an array called y_score. Use the predict method and pass it to the X_test dataset:

```
y_score = housevalue_regressor.predict(X_test)
```

You now have a vector, y_score, with the model's prediction of the housing value based on the inputs in X_test. You'll use y_score in just a bit to check the accuracy of the model by comparing y_score to the *ground truth* y_test values.

As before with the iris data in *Chapter 2*, if you want to predict a housing value based on data that is not in the test or training dataset, to do inference, you just pass the model the equivalent of one row of input data, and it will predict the housing value. Keep in mind that you must maintain the units used in the input data, so you need to write median income in terms of $10,000, and so on.

2. **Make a prediction (inference) based on example measurements**: Let's use the model to make a prediction and try this with a made-up example. To use the model to do inference, you need to provide data for each of the X columns, which are: median income, house age, average number of rooms for the block where the house is located, average number of bedrooms for the block, population of the block, the average occupancy, and the latitude and longitude of the block. For example, let's say the median income is $125,000, which is 12.5 in terms of $10,000. The house was built in 1980, so in 1990, it was 10 years old. It has 9 rooms total and 3 bedrooms; the population of the block where the house is located is 4000, and it's within Marina Del Ray, California, which has a latitude of 33.98 and a longitude of -118.45. The input array for this information looks like this:

MedInc	HouseAge	AveRooms	AveBedrms	Population	AveOccup	Latitude	Longitude
12.5	10	9	3	4000	3	33.98	-118.45

Figure 4.7 – Values for our example prediction

To use these values, first, create an `np.array` instance called `X_example` and reshape it to be one column with eight rows with `X_example = X_example.reshape(1,8)`. The input data is now in the format the model expects, and you can predict the associated house value using the `predict` method and passing in `X_example`. Assign that value to `y_example`:

```
X_example = np.array([12.5,10,9,3,4000,3,33.98,-118.45])
X_example = X_example.reshape(1,8)
y_example = housevalue_regressor.predict(X_example)
```

You now have the predicted value for a house with the properties from the example in `y_example`. Take a look at the answer by printing it out:

```
print(y_example)
```

The following is the output from printing `y_example`:

```
[5.3823247]
```

That's all you need to do to use the model. Your model predicts a value of 5.3823247, which is given in hundreds of thousands of US dollars ($100,000); in other words, $538,232.47. This makes sense for a larger home (nine rooms, three bedrooms) in a wealthier area (median income 12.5, which is given in thousands of US dollars = $125,000).

Next, let's look at how well the predictions it makes match the actual data. This is the equivalent of the confusion matrix you did for the classification model in *Chapter 2*. This time, you have a regression model, so use parameters that measure model fit, such as **R²** and RMSE.

3. **Test the effectiveness of the model by calculating the fit parameter R2 value**: Use the `r2_score` Scikit-learn function and pass it `y_test` as the `y_true` true y values, otherwise known as the ground truth, and `y_score` as the `y_pred` predicted values. Print the R^2 value to see the result:

```
predicter_r2 = r2_score(y_true=y_test, y_pred=y_score)
print(predicter_r2)
```

This gives the following output:

```
0.8186237017741059
```

The R^2 for our model is 0.8186, which is quite good. An R^2 of 1.0 would mean every predicted value is exactly the same as the actuals. This is very unlikely to occur. A very high R^2 value can be the result of overfitting, which will make the model less accurate for data other than the training data; in other words, when doing inference.

4. **Test the effectiveness of the model by calculating the fit parameter RMSE value**: Scikit-learn has a built-in function for **mean squared error** (MSE) called `mean_squared_error`. Use the `mean_squared_error` Scikit-learn function and pass it `y_test` as the `y_true` true y values, otherwise known as the ground truth, and `y_score` as the `y_pred` and use `squared=False` predicted values to get `predicter_rmse`. Print `predicter_rmse` to see the result:

```
predicter_rmse = mean_squared_error(
    y_true=y_test, y_pred=y_score, squared=False
)
print (predicter_rmse)
```

This gives the following result:

```
0.48699478645182126
```

The RMSE is on a scale from zero to one, with a lower RMSE being better. This RMSE is in the middle, so you should investigate. To do this, plot your model and your actual values, and also then plot the residuals – the part of the actuals that the model doesn't explain.

5. Investigate a regression model performance by plotting the model versus actuals, and the residuals: to start, plot the predicted y values, the target on the *y* axis, and the true *y* values that were in the test dataset on the *x* axis. Seaborn has a built-in function, `sns.regplot`, that fits a linear model and displays the line of fit. This can show you where your model is off from the true values. We set colors for the points with `scatter_kws=("color":"grey")` and the line with `line_kws=("color":"black")` in the call to `regplot`. Otherwise, they are the same color and hard to differentiate:

```
sns.regplot(
    x=y_test, y=y_score,
    scatter_kws={"color": "grey"},
```

```
        line_kws={"color": "black"}
    )
```

This produces the following graph:

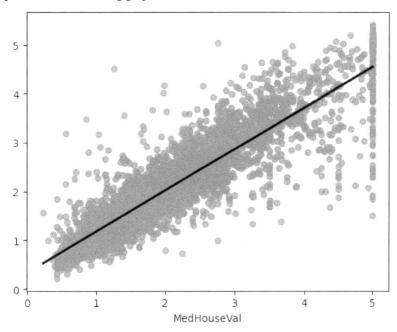

Figure 4.8 – Scatter plot with the actual median house value in the test dataset on the x axis
and the predicted median house value from our model using the test dataset on the y axis

In the graph you made, you have a pretty good fit at the low end of the *x* axis, but at higher values, your model doesn't work as well, especially since there is a hard limit in the actual data for house value. You can look at the **residuals** to understand more. A residual is the part of the actual value that is not explained by the model. Residuals are calculated by subtracting the fitted value from the observed value. You can do this calculation by simply subtracting, as seen here:

```
residuals = y_test - y_score
```

Take a look at the values by printing residuals:

```
print (residuals)
```

This results in the following output:

```
18403    -0.225832
13405     0.042021
9539     -0.249945
8668      0.515881
9019     -0.470594
          . . .
10696    -0.058746
20360    -0.252144
10281     0.487568
7092     -0.213285
3219      0.058771
Name: MedHouseVal, Length: 4128, dtype: float64
```

Figure 4.9 – Results from printing the residuals

Ideally, we have residuals that are very close to zero and both above and below zero. To more easily check this, plot the residuals against each of the input data columns.

6. To get ready to make this visualization, you need the actual X values and the residuals in a single DataFrame. Make this DataFrame by concatenating X_test and residuals. We chose to call the result X_testResiduals. Look at the result with the .head() method:

```
X_testResiduals = pd.concat([X_test, residuals], axis=1)
X_testResiduals.head()
```

This puts the X_test values and residuals into a single DataFrame so that we can build the visualization of the residuals against every input variable. It looks like this:

	MedInc	HouseAge	AveRooms	AveBedrms	Population	AveOccup	Latitude	Longitude	MedHouseVal
18403	5.4741	24.0	6.587799	1.060181	4017.0	3.311624	37.24	-121.84	0.225832
13405	1.9583	7.0	5.362805	1.442073	1064.0	3.243902	34.10	-117.46	0.042021
9539	2.6058	40.0	3.885714	0.914286	785.0	4.485714	37.39	-120.71	-0.249945
8668	4.5817	30.0	4.146135	1.108696	1526.0	1.842995	33.82	-118.39	0.515881
9019	8.0137	9.0	7.734673	1.041211	12873.0	3.274739	34.16	-118.78	-0.470594

Figure 4.10 – Results from X_testResiduals.head()

Now, the data is all together, and you can make a visualization to look for any correlations between the residuals and the input variables.

7. Use matplotlib.pyplot (plt) for this visualization because it has a feature where you can create an array of subplots and then direct output to each of those subplots in turn. Start by setting up a two-by-four set of subplots using fig, axes = plt.subplots(nrows=2, ncols=4, figsize=(24,16), sharey=True). By trial and error, we determined to set the size of the graphs to 24x16 using figsize=(24,16), sharey=True, which causes

the same *y*-axis scale to be used for all the graphs. This allows you to easily compare how the residual looks against each X parameter. Then, for each parameter in X_testResiduals, make an *x-y* scatter plot with y=X_testResiduals["MedHouseVal"].alpha = 0.5 makes the points slightly transparent so that it's easier to see overlapping data; color = "grey" sets the color for the scatter plot points:

```
fig, axes = plt.subplots(
    nrows=2, ncols=4, figsize=(24,16), sharey=True)
axes[0,0].scatter(x=X_testResiduals["MedInc"],
    y=X_testResiduals["MedHouseVal"],
    alpha=0.5, color="grey")
axes[0,0].set_title("MedInc")
axes[0,1].scatter(x=X_testResiduals["HouseAge"],
    y=X_testResiduals["MedHouseVal"],
    alpha=0.5, color="grey")
axes[0,1].set_title("HouseAge")
axes[0,2].scatter(x=X_testResiduals["AveRooms"],
    y=X_testResiduals["MedHouseVal"],
    alpha=0.5, color="grey")
axes[0,2].set_title("AveRooms")
axes[0,3].scatter(x=X_testResiduals["AveBedrms"],
    y=X_testResiduals["MedHouseVal"],
    alpha=0.5, color="grey")
axes[0,3].set_title("AveBedrms")
axes[1,0].scatter(x=X_testResiduals["Population"],
    y=X_testResiduals["MedHouseVal"],
    alpha=0.5, color="grey")
axes[1,0].set_title("Population")
axes[1,1].scatter(x=X_testResiduals["AveOccup"],
    y=X_testResiduals["MedHouseVal"],
    alpha=0.5, color="grey")
axes[1,1].set_title("AveOccup")
axes[1,2].scatter(x=X_testResiduals["Latitude"],
    y=X_testResiduals["MedHouseVal"],
    alpha=0.5, color="grey")
axes[1,2].set_title("Latitude")
axes[1,3].scatter(x=X_testResiduals["Longitude"],
    y=X_testResiduals["MedHouseVal"],
    alpha=0.5, color="grey")
axes[1,3].set_title("Longitude")
```

This will print out the following plot:

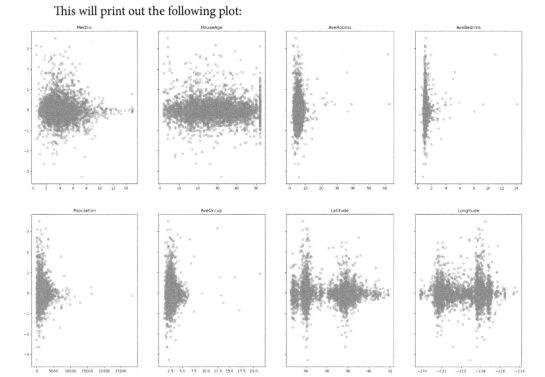

Figure 4.11 – Plots of the residual median house value versus each of the X parameters

This produces a grid of eight *x-y* scatter plots with the residual on the *y* axis and each input parameter in its own graph on the *x* axis. The residual median house value looks randomly distributed with respect to each of the X parameters. This is what you want to see: a normal distribution of the residual around zero. If you see patterns or curvature in the residual, then you know your model is incorrect and there is bias in the predicted values. If you see this, then you need to do some feature engineering and re-fit your model. You can read more about handling data problems in *Chapter 6*, and about feature engineering in *Chapter 7*.

You can only make predictions for parameters that are in the dataset

Notice that you are not using the data to predict the *selling price of a house*. That information is not available in the dataset, so you have no way to build a model for that parameter, given the data you have. If you want to know the house selling price rather than house value, you would need to either collect that information or use a different dataset that has that parameter.

You *can* do calculations based on information you have in the dataset to create input parameters. For example, you can divide the population by the average occupancy to get the average number of housing units in a block. This is called feature engineering. This will be covered more in *Chapter 7*.

In this section, you trained and tested a regression model that can predict the value of a house based on median income, house age, the average number of rooms for the block where the house is located, the average number of bedrooms for the block, population of the block, the average occupancy, and the latitude and longitude of the block. You used the model to do inference on example data and for a test data set created during the train/test split process. You looked at how well the model fits the test data by looking at R^2 and RMSE. You investigated the fit of the model by plotting the predicted y_score values against the y_test ground truth. Finally, you plotted the residual, the difference between y_score and y_test, versus each of the input parameters to look for patterns.

Comparing model parameters for the Iris and housing datasets

The key difference between the Iris dataset and this housing dataset is that the target parameter in the Iris dataset is a category, whereas in the housing dataset, it is a numeric parameter. Predicting a category is a classification problem, and in this chapter, you did a regression problem. That meant using a different method within XGBoost. This table compares the XGBoost training and inference calls used in *Chapter 2* and in this chapter:

	Iris dataset classification from *Chapter 2*	California housing dataset regression from *Chapter 4*
Set up the XGBoost model	`model = xgb.XGBClassifier(eval_metric="auc")`	`model=xgb.XGBRegressor()`
Train the XGBoost model	`model.fit(X_train,y_train, eval_set=[(X_test,y_test),(X_train,y_train)])`	`model.fit(X_train,y_train, eval_set=[(X_test,y_test),(X_train,y_train)])`
Use XGBoost for inference	`y_score = model.predict(X_test)`	`y_score = model.predict(X_test)`

Figure 4.12 – Comparing model parameters for the Iris and housing datasets

This table compares the model parameters for the Iris and housing datasets. The major difference is that the Iris dataset is a classification problem and the housing dataset is a regression problem, so the code used to set up XGBoost to model the data is different. For the iris data, you called XGBClassifier, and for the housing data, you used XGBRegressor. The evaluation metric is also different; for classification, it is **Area Under the Curve** (**AUC**), whereas in regression, you used the default of RMSE when you trained the model. The code to train XGBoost is the same in both cases. The code to predict values is also the same.

In this chapter and *Chapter 2*, you explored the data by making graphs before building a model, and you prepared the data for modeling by splitting it into a training dataset and a testing dataset. The way you evaluated the models was different between the two chapters. For the iris data, you used accuracy, precision, recall, and F1 score to test the model's effectiveness. These are classification metrics. Here, you used R2 and RMSE, which are regression metrics.

Summary

This chapter built upon the work you did in *Chapter 2*. You changed out the dataset, switching the target from species of irises, which is categorical data, to housing value data, which is a continuous numeric variable. You built a regression model to predict house value given the various **features** (X parameters). As you did in *Chapter 2*, you built graphs to look at the data before building a model. You prepared the data to create test and training datasets and trained a regression model with XGBoost. Because it's a regression model, this time, you used R^2, RMSE, and residuals to evaluate the model.

So far, you've used XGBoost to build a classification model and, in this chapter, a regression model. The regression tree model created is a gradient-boosted tree model because you used the default settings in XGBoost. XGBoost is an ensemble tree method, and with it, you can build both gradient-boosted tree models and random forest models. The next chapter, *Chapter 5*, will discuss which type of model is best for your data, as well as when using **neural networks** (NNs) and **deep learning** (DL) is more appropriate.

Part 2: Practical Applications – Data, Features, and Hyperparameters

In this part, you will explore solutions to common questions that arise when modeling data. Through guided activities, you will learn when to use ensemble models or simple classification and regression tree models. You will work with model metrics and apply methods to address common problems with real-life datasets. By applying feature engineering methods, you will improve model performance and handle text and time-based datasets. This part ends with an exploration of interpreting XGBoost models and some practice in extracting feature importance.

This part contains the following chapters:

- *Chapter 5, Classification and Regression Trees, Ensembles, and Deep Learning Models – What's Best for Your Data?*

- *Chapter 6, Data Cleaning, Imbalanced Data, and Other Data Problems*

- *Chapter 7, Feature Engineering*

- *Chapter 8, Encoding Techniques for Categorical Features*

- *Chapter 9, Using XGBoost for Time Series Forecasting*

- *Chapter 10, Model Interpretability, Explainability, and Feature Importance with XGBoost*

Classification and Regression Trees, Ensembles, and Deep Learning Models – What's Best for Your Data?

This chapter covers when to use ensemble models or simple classification and regression tree models. Through experimenting with the housing dataset that we used in *Chapter 4*, you will see the differences between models in terms of model accuracy, as measured by R^2 and RMSE, how long it takes for the models to be trained, and how long it takes for the models to make predictions. To do so, you will test and compare XGBoost, scikit-learn gradient boosting, and random forest models. You will learn when to use decision-tree-based machine learning versus deep learning. Lastly, you will learn how to set the various hyperparameters (parameters that control how the model learns) for XGBoost.

In this chapter, we will cover the following main topics:

- When to use ensemble models versus single **Classification and Regression Tree** (**CART**) models
- Comparing models with the housing dataset
- Comparing XGBoost to linear regression
- Comparing XGBoost to CART
- Comparing XGBoost to gradient boosting and random forest
- When to use decision-tree-based models versus deep learning models
- Setting XGBoost regression hyperparameters
- Results from comparing models with the housing dataset

Technical requirements

This chapter will build on the work you did in *Chapter 4*. You can use the same virtual environment that you set up with the necessary packages previously. The code presented in this chapter is available in this book's GitHub repository at `https://github.com/PacktPublishing/XGBoost-for-Regression-Predictive-Modeling-and-Time-Series-Analysis`.

You will use the following software:

- Python 3.9
- XGBoost 1.7.3
- NumPy 1.21.5
- pandas 1.4.2
- scikit-learn 1.4.2
- Seaborn
- Anaconda
- VS Code

When to use ensemble models versus single CART models

XGBoost is an **ensemble model**, where multiple CART decision trees are created and the results are aggregated. You learned about CART models in *Chapter 3*. In this chapter, you will work on predictions using the housing dataset, which is the same data that you used in *Chapter 4*. You may want to start from your code in *Chapter 4* and modify it as you experiment with the various types of models.

By combining the results from multiple trees to generate the output, ensemble models have smaller residuals and improved R^2 and RMSE values than simple tree models. The risk of using ensemble models is overfitting. Models that have overfitting work well on the training dataset but do not provide as good a fit on other data because they are too specialized. This is a problem because your model will be inflexible and predictions made on data that isn't *exactly* like your training data will be less accurate. What this means in practice is that a model with a less perfect RMSE will perform better than one that has a very low RMSE due to overfitting. You cannot see overfitting by just looking at the RMSE or R^2 values; you need to plot the points and the line of fit, as shown in the following figure, or plot the residuals, as you did in *Chapter 4*:

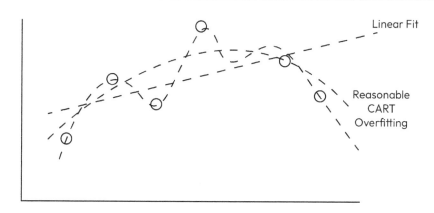

Figure 5.1 – Illustration of a poor fit, a reasonable fit, and overfitting

In this graph (*Figure 5.1*), the circles indicate data in the training dataset. The straight dotted line is a linear fit and has the highest RMSE of the three fit lines shown. The curved dotted line is a reasonable fit from a CART model. The solid line has the least RMSE as it passes exactly through each point; this is known as **overfitting**. While the model with the least RMSE works well for the training data, any other dataset provided will produce poor results. The model is inflexible.

It is good practice to start with the simplest model and then work toward a more complex model if your RMSE does not meet your criteria. For example, in *Figure 5.1*, we started with a linear fit, and you can see that there is a large space – a residual – between the line of fit and some points. Thus, it makes sense to try a CART model on the data. Then, looking at the residuals, you can determine if a more complex model is needed, such as an ensemble model. Let's put this into practice by looking at the housing dataset.

Comparing models with the housing dataset

Let's start by loading and preparing the data for modeling:

1. **Set up the Python environment**: Start by making a copy of the code from *Chapter 4* and modifying it. Like in *Chapter 4*, we'll need the pandas and NumPy libraries. We've chosen to name the file housingvaluemodelcomparison.ipynb:

    ```
    # ---------------------------------------
    # filename housingvaluemodelcomparison.ipynb
    # purpose compare predictions of house value
    # by different models
    # author Joyce Weiner
    # revision 1.0
    # revision history 1.0 - initial script
    # ---------------------------------------
    ```

```
import pandas as pd
import numpy as np
```

2. **Load the California housing dataset from scikit-learn**: The housing dataset is built into scikit-learn. The `fetch_california_housing` function has two useful parameters that allow you to load the data as a pandas DataFrame and put the features (X values) and the target y value into separate variables in just one line of code:

```
# load the California Housing dataset from scikit-learn
from sklearn import datasets
housingX, housingy = datasets.fetch_california_housing(
    return_X_y=True, as_frame=True)
```

With that, you have two DataFrames – one called `housingX`, which contains the feature data, and one called `housingy`, which contains the target values.

3. **Prepare the data for modeling**: To start preparing to make a model, you must split our data into testing and training sets using the `train_test_split` function from scikit-learn. You can reuse the code from *Chapter 4* to do this. You'll put both X and y in one line and provide four variable names to store the results. We've kept the variable names the same as they were in *Chapter 4*: `X_train`, `X_test`, `y_train`, and `y_test`. As before, you'll have 80% of the data in the training set, reserving 20% for testing the models. You can set this with `test_size=0.2`. Here, `random_state=17` is an arbitrary value that will seed the random selection of which columns go into which set. By using the same value as we do here, you'll have the same rows in your data and our outputs will match:

```
from sklearn.model_selection import train_test_split
X_train, X_test, y_train, y_test = train_test_split (
    housingX, housingy, test_size=0.2, random_state=17)
```

At this point, you can do additional data cleaning. Note that this isn't needed here because you're using an example dataset.

With that, you've loaded and prepared the data for modeling. Next, we'll build multiple models on the same dataset so that you can compare how the different algorithms perform.

Comparing XGBoost to linear regression

When building models, you want to start with the simplest and increase the complexity of the model only if needed. This approach means your models will be fast, small, and easy to explain, and only take more compute resources and training time if the simple options don't work. Taking this same approach for these examples, start by comparing XGBoost to a plain linear fit. Recall from *Chapter 4* that your XGBoost model had an RMSE of 0.487 and an R^2 of 0.819. You're looking to match these values or better. RMSE measures the amount of error between the predicted values and the known true values. Therefore, s better RMSE is lower, meaning there is less error. A better R² is higher, meaning the model fits the data better. Let's get started:

1. **Perform a linear fit on the data**: The easy way to perform linear regression is to use scikit-learn, which contains a linear regression model. Note that you don't need to train it, though to get a fair comparison, you should use the training dataset. Pass `LinearRegression.fit()` the matrix of X values, `X_train`, and the matching y values, `y_train`. scikit-learn uses ordinary least squares or non-negative least squares with a wrapper to return a predictor object matching the output from other models in the package. We've added `%%time` at the beginning of the cell to measure execution time:

   ```
   %%time
   from sklearn.linear_model import LinearRegression
   housing_linear_regression = LinearRegression().fit (
       X_train, y_train)
   ```

   When you run the cell, the following times will be shown. Your results might be different:

   ```
   CPU time: total: 31.2 ms
   Wall time: 2.71 s
   ```

   This took 2.71 seconds of wall time and 31.2 **milliseconds (ms)** of CPU time on our computer. Wall time is the total elapsed time from the start to the end of cell execution, while CPU time is how long the CPU spent executing the cell. Wall time includes memory or storage access time and can vary if your computer is busy with concurrent tasks. You'll want to keep track of both the CPU time and the wall time needed to train or fit the models as you go through this chapter so that you can compare how they perform against each other.

2. **Check the RMSE for the linear model**: To check the RMSE, you can use the `root_mean_squared_error` function from scikit-learn. First, call the `predict` method for our linear regression model on your test data, `X_test`, to predict output values. Then, compare the predicted values to the true, actual `y_test` values using the `root_mean_squared_error` function.

   ```
   from sklearn.metrics import root_mean_squared_error
   housing_linreg_ypred = housing_linear_regression.predict(X_test)

   housing_linreg_rmse = root_mean_squared_error(y_true=y_test,
       y_pred=housing_linreg_ypred)
   print ("Linear regression RMSE is {0:.2f}".format(
       housing_linreg_rmse))
   ```

   You'll see the following output:

   ```
   Linear regression RMSE is 0.72
   ```

   This is a much larger RMSE than you had with XGBoost in *Chapter 4*. What this is saying is that this model isn't very good at predicting the housing value.

3. **Check the R^2 value for the linear model**: To check the R^2 value, scikit-learn has the `.score` method, which you can call to get the R^2 value for a fit. The first line uses `.score` to get the R^2 value for the `housing_linear_regression` model for the test dataset. The second line outputs the R^2 value formatted to show only two decimal places, using `{0:.2f}` and `.format(linr2)`, just as we did with RMSE. We're storing all the RMSE and R^2 values from the various models in variables so that you can make a list of them at the end programmatically and compare how each model performed:

```
housing_linreg_r2 = housing_linear_regression.score(X_test,
    y_test)
print("Linear regression Rsquared is {0:.2f}".format(
    housing_linreg_r2))
```

This produces the following result:

```
Linear regression Rsquared is 0.60
```

The R^2 value is only 0.60, which is not as good as the R^2 value we got with XGBoost. This is the measurement of how well the model fits the data. The higher the number, the better.

4. **Plot the predicted values compared to the actuals**: To wrap up the linear fit section, plot the predicted values versus the actuals for the test dataset to get a picture of how well this model works for your data. You can use Seaborn to make the plot, just as you did in *Chapter 4*, to look at the results of the XGBoost prediction. You can use `regplot`, which fits a linear model and displays the line of fit to compare the actual y values (`x=y_test`) on the *x* axis to the predicted y values on the *y* axis (`y=housing_linreg_ypred`). The `scatter_kws=("color":"grey")` and `line_kws=("color":"black")` lines set the colors for the points to gray and the line of fit to black; otherwise, they will be the same color and hard to differentiate:

```
import seaborn as sns
sns.regplot(x=y_test, y=housing_linreg_ypred,
    scatter_kws={"color": "grey"}, line_kws={"color": "black"})
```

Here's the output:

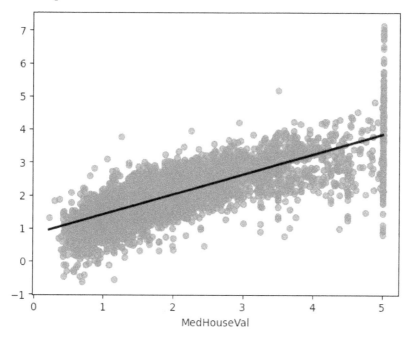

Figure 5.2 – Linear regression scatterplot

This scatterplot shows the actual median house value on the *x* axis and the prediction from the linear regression on the *y* axis. By visually comparing the line of fit to the points, you can see a curvature at higher actual median house values. This means that the fit isn't good in that region.

Now, it's time to try another model. But before you do that, you will learn about when to use linear regression.

When to use linear regression

So far, XGBoost is doing quite well. If the linear regression RMSE was low and the R^2 value was high, say 0.8 or better, then you might want to use linear regression rather than a more complex ensemble model. First of all, linear regression is a quick model to fit, meaning that it will be easy to maintain when it's put to use. Secondly, it's easy to explain why the model is making the prediction; you can retrieve the equation for the linear fit using the `.intercept_` and `.coef_` attributes and share the equation that is being used to calculate y:

$$y = \left(coef_1 \times x_1 + coef_2 \times x_2 + \cdots \; coef_n \times x_n \right) + intercept$$

Next, you will compare XGBoost to a regression tree (CART) model.

Comparing XGBoost to CART

So far, you've compared XGBoost to linear fit. For the housing data, you've seen that XGBoost provides an improvement with better RMSE and R² values, even though it takes longer to execute than linear regression. In the following steps, we will compare it to CART in terms of fit and performance, as implemented by scikit-learn:

1. **Fit a regression tree model using CART**: To fit a CART model, use the DecisionTree module in scikit-learn, specifically `sklearn.tree.DecisionTreeRegressor` since you will be performing regression to predict housing values. First, set up the model by calling `DecisionTreeRegressor` and put the result in `housing_CART`. Use the default settings so that no values are passed. This is where you can set the model hyperparameters. Next, fit the model by applying the `.fit` method:

    ```
    %%time
    from sklearn.tree import DecisionTreeRegressor
    housing_CART = DecisionTreeRegressor()
    housing_CART_regression = housing_CART.fit(X_train, y_train)
    ```

    This results in a trained or fit regression model using the default settings from scikit-learn. On our computer, this took 218 ms of wall time and 141 ms of CPU time. The default settings in scikit-learn for `DecisionTreeRegressor` are shown in the following table (*Table 5.1*):

Parameter	Default Value	What It Configures
criterion	squared_error	A function that measures the quality of a split.
splitter	best	Select the best split rather than randomly selecting a split.
max_depth	none	How deep a tree to build during training in terms of the number of layers.
min_samples_split	2	The minimum number of samples required to allow a split.
min_samples_leaf	1	The minimum number of samples in a leaf node.
min_weight_fraction_leaf	0.0	Defines the weighting required at a leaf node.
max_features	none	The number of features to consider when selecting a split. When set to none, all features are used.
random_state	none	Controls the randomness of each split.

Parameter	Default Value	What It Configures
`max_leaf_nodes`	`none`	Here, `none` allows an unlimited number of leaf nodes.
`min_impurity_decrease`	`0.0`	A node will split if the impurity is greater or equal to this threshold.
`ccp_alpha`	`0.0`	A complexity parameter that's used for minimal cost-complexity pruning.

Table 5.1 – Default parameters for DecisionTreeRegressor

2. **Predict**: Make a prediction using the `predict` method and pass in the `X_test` data:

    ```
    %%time
    housing_cart_ypred = housing_CART_regression.predict(X_test)
    ```

    At this point, you have predictions from the test dataset, `X_test`, stored in `housing_cart_ypred` that you will compare with `y_test` and calculate the necessary RMSE and R^2 values. Making the prediction took 0 ms of CPU time – in other words, it was very fast. On our computer, it took 3.99 ms of wall time.

3. **Check the RMSE value for the CART model**: You can use `root_mean_squared_error` and set `y_true=y_test` and `y_pred=housing_cart_ypred` to return the RMSE. We put the result in `housing_cart_rmse`:

    ```
    housing_cart_rmse = root_mean_squared_error(y_true=y_test,
        y_pred=housing_cart_ypred)
    print ("CART RMSE is {0:.2f}".format(housing_cart_rmse))
    ```

    This prints out the following result:

    CART RMSE is 0.72

    While this RMSE is not as good as XGBoost, it is the same as the linear model. Very interesting! Next, we'll compare the R^2 value.

4. **Calculate the R^2 value for the CART model**: Use the `.score` method and pass it `X_test` and `y_test` to get the R^2 value. Store the result in `housing_cart_r2`:

    ```
    housing_cart_r2 = housing_CART_regression.score(X_test, y_test)
    print("Rsquared is {0:.2f}".format(housing_cart_r2))
    ```

    This results in the following output:

    CART Rsquared is 0.60

Just like the RMSE, the R² is the same as you got from the linear fit. These results are very interesting. The values, to two decimal places, match what you got for linear regression. Sometimes, a more complex model isn't as effective, while other times, an ensemble model will enable improved fit. We'll cover this shortly. First, though, let's plot the predicted y values against the true y values and see how well the CART model has fit the data and if there is a difference in the plot compared to the linear fit.

5. **Plot the predicted values compared to the actuals**: Pass the regplot function the x=y_test actuals to get the x values and your predictions from the CART model as the y values via y=housing_cart_ypred. Set the colors with scatter_kws=("color":"grey") and line_kws=("color":"black") so that they aren't the same and you can see the line versus the points:

    ```
    sns.regplot(x=y_test, y=housing_cart_ypred,
        scatter_kws={"color": "grey"}, line_kws={"color": "black"})
    ```

This results in the following graph:

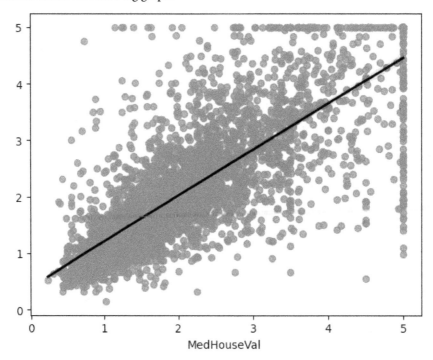

Figure 5.3 – Scatterplot comparing the CART prediction to actuals

This scatterplot (*Figure 5.3*) shows the median house value in the test dataset on the *x* axis and the predicted median house value from our CART regression when using the test dataset on the *y* axis. It looks very different than the results we got when performing linear regression. Visually, you can see that the CART model performs better; the points are more centered in the graph along the diagonal. There *are* still aspects of the actual data that the model is not predicting; if it was, all the points would be on the diagonal line of x=y. With that, you've seen how well CART does when it comes to modeling the data and comparing it to XGBoost. CART is faster, but it doesn't fit the data as well.

When to use CART

As you've seen, CART is an improvement over linear regression when there are non-linear terms in the data that need to be included in the model for an accurate prediction. The downside is that the resulting model is not as easy to explain to someone else. Linear regression can be explained by sharing a simple equation relating the inputs to the output. While CART is more challenging to explain than linear regression, CART is easier to explain than an ensemble model such as XGBoost since it only builds one decision tree. You can show the CART decision tree you built by calling `tree.plot_tree` and passing the regression model – that is, `housing_CART`:

```
from sklearn import tree
tree.plot_tree(housing_CART)
```

This results in the following plot:

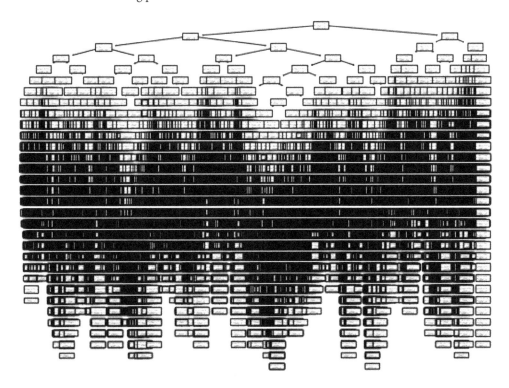

Figure 5.4 – Plot of the regression tree created by CART

The intention behind including this figure is for you to get a sense of the overall intricacy of the model, not for you to squint to attempt to read the values in the nodes. The default settings for CART have created a fairly complicated decision tree with a huge number of layers. There is a risk of overfitting due to such a complex model. This can be controlled with the max_depth setting. Another technique to consider is pruning a tree back once it has been generated. You will learn about both of these settings for XGBoost later in this chapter.

Next, you will compare XGBoost to ensemble tree methods, gradient boosting trees, and random forest models.

Comparing XGBoost to gradient boosting and random forest models

In this section, you will build a gradient-boosted tree model and a random forest model with scikit-learn and compare them to XGBoost. Gradient boosting and random forest are types of **ensemble** tree-based models. Recall from *Chapter 3* that the authors of XGBoost made several improvements to speed up execution, but many of the changes they made were for scalability so that the algorithm could handle large datasets effectively. What does the regular gradient boosting tree model do on this small example dataset? In this section, you will build a gradient boosting model and a random forest model using scikit-learn. scikit-learn is consistent in how models are called and fit, so the steps here will be similar to what you have just completed for CART. Let's get started:

1. **Fit a regression model using gradient-boosted trees**: Import GradientBoostingRegressor from sklearn.ensemble, which contains the ensemble models. Create a GradientBoostingRegressor value and call it housing_gbt. Set random_state = 17 to ensure you get the same results as in this book. Then, set max_depth = 6 so that it matches the XGBoost default tree depth. Then, fit the model by calling the .fit method and passing in X_train and y_train:

    ```
    %%time
    from sklearn.ensemble import GradientBoostingRegressor
    housing_gbt = GradientBoostingRegressor(random_state=17,
        max_depth=6)
    housing_gbt_regression = housing_gbt.fit(X_train, y_train)
    ```

    This results in a trained gradient-boosted regression model that uses the default settings from scikit-learn, except for max_depth, which you changed to match the XGBoost default. On our computer, this took 10,300 ms of wall time and 5,840 ms of CPU time. This was much slower than a non-ensemble CART model, which makes sense since this model had to build multiple decision trees rather than just one. The default setting for scikit-learn is to build 100 boosting stages. The default tree depth is max_depth = 3. This setting is one you may want to experiment with to see if you can improve the fit metrics for the model.

2. **Predict the results for the X_test dataset**: Call the .predict method and pass X_test. Put the results in housing_gbt_ypred:

```
%%time
housing_gbt_ypred = housing_gbt_regression.predict(X_test)
```

Now, you have the predicted y values based on the X_test dataset stored in housing_gpt_ypred. You will use these to calculate the fit metrics. As with CART, the prediction calculation is fast, with 15.6 ms of CPU time and 24.9 ms of wall time. Next, we will look at how well this model performs.

3. **Calculate the RMSE value for the predicted y values**: You can use root_mean_squared_error and set y_true=y_test and y_pred=housing_gbt_ypred to return the RMSE value. Put the result in housing_gbt_rmse:

```
housing_gbt_rmse = root_mean_squared_error(y_true=y_test,
    y_pred=housing_gbt_ypred)
print ("Gradient boosting regressor RMSE is {0:.2f}".
format(housing_gbt_rmse))
```

This results in the following output:

Gradient boosting regressor RMSE is 0.49

This is much improved compared to the CART RMSE and matches the result from XGBoost. Next, we'll check the R^2 value.

4. **Calculate the R^2 value for the predicted y values**: You can use the .score method and pass it X_test and y_test to get the R^2 value. Store the result in housing_gbt_r2:

```
housing_gbt_r2 = housing_gbt_regression.score(X_test, y_test)
print("Gradient boosting regressor Rsquared is {0:.2f}".
format(housing_gbt_r2))
```

This prints out the following:

Gradient boosting regressor Rsquared is 0.82

The R^2 value has also improved compared to CART and matches the result from XGBoost. Let's plot the predicted y values against the true y values and see where the gradient boosting model has improved the fit compared to CART.

5. **Plot the predicted values compared to the actuals**: Pass the regplot function the x=y_test actuals as the x values and your predictions from the gradient boosting model as the y values via y=housing_gbt_ypred. Set the colors with scatter_kws=("color":"grey") and line_kws=("color":"black") so that they're not the same:

```
sns.regplot(x=y_test, y=housing_gbt_ypred,
    scatter_kws={"color": "grey"}, line_kws={"color": "black"})
```

This outputs a graph:

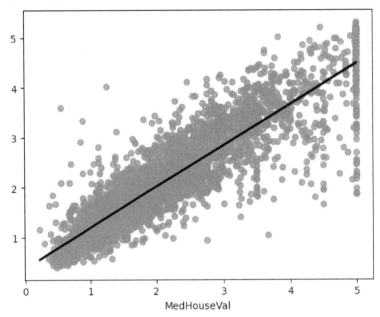

Figure 5.5 – Scatterplot comparing the gradient boosting prediction to the actuals

This scatterplot shows the median house value in the test dataset on the x axis and the predicted median house value from your gradient boosting regression using the test dataset on the y axis. There is a clear tightening of the predictions from the gradient boosting model compared to the predictions from CART. This is expected given the improvement in the RMSE and R^2 values.

With that, you've compared a classic gradient-boosted tree algorithm found in scikit-learn with XGBoost. Next, you will compare XGBoost to scikit-learn's random forest algorithm.

6. **Fit a regression model using random forest**: Import RandomForestRegressor from sklearn.ensemble. Then, build a model called housing_rf and set random_state=17 so that it matches the results in this book. Train a model called housing_rf_regression using the .fit method and passing in the X_train and y_train datasets:

```
%%time
from sklearn.ensemble import RandomForestRegressor
housing_rf = RandomForestRegressor(random_state=17)
housing_rf_regression = housing_rf.fit(X_train, y_train)
```

This results in a trained random forest regression model. The model training took 13,400 ms and 19,800 ms. This is quite a lot longer than for gradient boosting. The extra time might be worth it if the model results in improved fit metrics. You will prepare for that next by predicting values for y based on the X_test data to be used in the fit metrics.

7. **Predict the results for the X_test dataset**: Call the `.predict` method for the trained model, `housing_rf_regression`, and pass in the `X_test` dataset. Name the result `housing_rf_ypred`:

```
%%time
housing_rf_ypred = housing_rf_regression.predict(X_test)
```

The predicted y values are now in `housing_rf_ypred`. This prediction took more time than any of the other models – that is, 62.5 ms of CPU time and 163 ms of wall time. This is because to calculate a prediction, the model must traverse all the trees in the forest. The more trees in the model, the longer it will take to calculate a prediction. This might be worth it if the fit metrics are improved. You will check those next, starting with the RMSE.

8. **Calculate the RMSE for the predicted y values**: Use `root_mean_squared_error` and pass `y_true = y_test`, `y_pred = housing_rf_ypred`. Put the result in `housing_rf_rmse`:

```
housing_rf_rmse = root_mean_squared_error(y_true=y_test,
    y_pred=housing_rf_ypred)
print ("Random Forest RMSE is {0:.2f}".format(
    housing_rf_rmse))
```

This prints out the following:

```
Random Forest RMSE is 0.51
```

The random forest model's RMSE is slightly worse at 0.51 than the gradient-boosting model's RMSE at 0.49. Next, you must calculate the R^2 value for the random forest model.

9. **Perform a calculation**: Calculate the R^2 for the predicted y values:

```
housing_rf_r2 = housing_rf_regression.score(X_test, y_test)
print("Random forest Rsquared is {0:.2f}".format
    (housing_gbt_r2))
```

This prints out the following:

```
Random forest Rsquared is 0.82
```

The random forest model's R^2 value is the same as it is for the gradient boosting model, even though the RMSE value is not quite as good. It matches XGBoost as well. Now, you must plot a graph to see how the predicted values compare to the actuals.

10. **Plot the predicted values compared to the actuals**: Use Seaborn's `regplot` with `y_test` on the *x* axis and `housing_rf_ypred` on the *y* axis. Change the colors with `scatter_kws={"color": "grey"}`, and `line_kws={"color": "black"})` so that the line of fit is visible:

```
sns.regplot(x=y_test, y=housing_rf_ypred,
    scatter_kws={"color": "grey"}, line_kws={"color": "black"})
```

This results in the following graph:

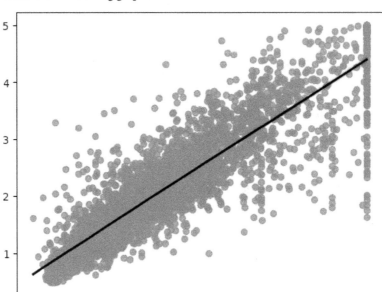

Figure 5.6 – Scatterplot comparing the random forest prediction to the actuals

This scatterplot shows the median house value in the test dataset on the *x* axis and the predicted median house value from your random forest regression using the test dataset on the *y* axis. There is a clear tightening of the predictions from the random forest model compared to the predictions from CART, which is as you would expect given the improvement in the RMSE and R^2 values. Compared to gradient boosting, the predictions are *not* as tight around the line of fit. For this dataset, we get better results from gradient boosting.

With that, you've compared XGBoost to scikit-learn's random forest algorithm and gradient boosting. You saw that XGBoost matches performance to scikit-learn's gradient boosting algorithm in terms of the overall training and prediction time for smaller datasets. You also saw how gradient boosting algorithms outperformed random forest models in terms of training time. Next, you will learn about when to use ensemble models in general.

When to use ensemble tree-based models

As you saw with this example, sometimes, a simpler modeling method does not provide enough accuracy or is unable to meet the criteria for fit metrics such as R^2 or RMSE. In this situation, or if you notice signs of higher-order patterns when you plot residuals (*Figure 5.7*), consider using ensemble methods:

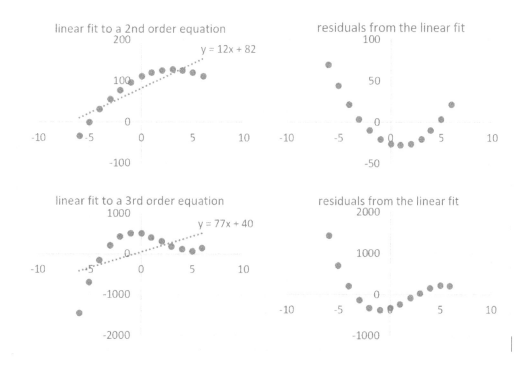

Figure 5.7 – Examples of residuals with signs of higher-order patterns

The graph at the top left of *Figure 5.7* plots the actual data and a linear fit for a second-order equation. The graph of the residuals on the right has curvature, indicating that there are higher-order terms. In this example, there's a second-order term that the fit does not capture. Similarly, the graph at the bottom left plots the actual data and a linear fit for a third-order equation and the resulting residuals. Again, there is curvature, indicating that the model accurately represents the data. If you see similar curvature in residuals, then an ensemble model may provide better results.

The disadvantages of ensemble methods are additional training time, reduced ability to explain how the model determines the predicted values, and increased potential for overfitting.

When to use simple models versus ensemble tree models

As you learned in this chapter, it is a good idea to start with a simple model and only add complexity if needed to improve the fit of the training data. Simple models provide three benefits:

- They are fast to train
- The predictions are easier to explain
- They are less prone to overfitting, which creates a model that works well only on the training data

Simple models and machine learning techniques such as CART and ensemble methods do not perform as well as deep learning models for computer vision tasks or speech recognition. Deep learning requires more training data and more training time than the methods you have used in this chapter.

This wraps up this section on ensemble tree models. In this section, you built and trained a gradient boosting predictor for the housing dataset using the default settings and the standard gradient boosting algorithm in scikit-learn. You also built and trained a random forest predictor for the housing dataset using scikit-learn. Then, you calculated the R^2 and RMSE values for the two models and compared them with XGBoost. By doing this, you saw that for a smaller dataset like the housing dataset, the standard gradient boosting regressor model in scikit-learn performed on par with XGBoost in terms of both fit metrics and training and prediction time. Next, you will learn when to choose decision-tree-based models such as XGBoost and when choosing a deep learning model is better.

When to use deep learning models versus decision-tree-based models

Certain applications benefit from the use of deep neural networks, called deep learning. In those instances, the neural network outperforms ensemble tree models. The following are some use cases for deep learning:

- Machine vision (for example, object recognition)
- Natural language understanding (for example, speech-to-text)
- Large language models (for example, generative AI)

Next, you will learn about the hyperparameters that can be used for XGBoost regression and how to set them.

Setting XGBoost regression hyperparameters

Hyperparameters are settings that affect model training. In this section, you will learn about the hyperparameters for XGBoost, what their default settings are, and how to change these settings. This section will cover regression in particular. Typically, you will use experimentation to decide on the best hyperparameter values to use when creating a model for the data. This is because the settings are dataset-dependent. There are software packages such as SigOpt that help keep track of your hyperparameter experiments and automate the process. These packages are outside the scope of this book.

The following table (*Table 5.2*) lists the hyperparameters you should consider modifying to suit your dataset. This is a subset of the available XGBoost learning parameters that are listed in the XGBoost documentation: `https://xgboost.readthedocs.io/en/stable/parameter.html`. XGBoost is highly configurable:

Setting (Alias)	Range	Default Value	Effect of the Setting
`eta` (`learning_rate`)	`[0,1]`	`0.3`	Shrinks the feature weights after each boosting step to make the boosting more conservative.
`gamma` (`min_split_loss`)	`[0,∞]`	`0`	Minimum loss reduction required to make a split on a leaf node. The larger it is, the more conservative it is.
`max_depth`	`[0,∞]`	`6`	The maximum depth of a tree; 0 indicates no limit. Increasing the tree depth impacts memory usage by XGBoost.
`min_child_weight`	`[0,∞]`	`1`	If a tree partition results in a leaf node with a sum of instance weight less than this value, the building process will stop partitioning. The larger it is, the more conservative it is.
`max_delta_step`	`[0,∞]`	`0`	The maximum delta we allow each leaf output to be. This can help if classes are extremely imbalanced. A 0 value means no constraint, while higher values make the update step more conservative.
`subsample`	`[0,1]`	`1`	Percentage to randomly sample the training data before growing trees to prevent overfitting. Subsampling occurs once per boosting iteration.
`sampling_method`	`uniform` `gradient_based`	`uniform`	The method that's used to sample the training instances. Here, `gradient_based` is only available when `tree_method` is `gpu_hist`.
`alpha` (`reg_alpha`)		`0`	Performs L1 regularization on the absolute value of weights to prevent overfitting. This works as a type of feature selection as it reduces some weights to zero.
`lambda` (`reg_lambda`)		`1`	Performs L2 regularization on the sum of squares of the weights to prevent overfitting by forcing weights to be small but not zero.

tree_method	auto approx exact hist gpu_hist	auto	A method for tree construction that's used by XGBoost. Here, auto uses a heuristic to set the method, exact is the exact greedy algorithm, approx is the approximate greedy algorithm you learned about in *Chapter 3*, hist is a faster histogram-optimized approximate algorithm, and gpu_hist is a GPU implementation of the hist algorithm. Note that approx, hist, and gpu_hist are supported for distributed training.
process_type	default update	default	Here, default is the normal boosting process that grows new trees. The update value starts with an existing model and updates the trees.

Table 5.2 – XGBoost tree booster regression hyperparameters

In this section, you learned about the settings for XGBoost that affect model training, known as hyperparameters. You learned what these settings control, what the default values are, and how you can change them to meet the needs of the data you are modeling. Next, let's summarize the results from all of the models you've used on the housing dataset.

Results from comparing models with the housing dataset

To finish comparing the models, let's look at the results of testing the different models by putting them into a table. The following table (*Table 5.3*) summarizes the results from the comparisons and looks at training time, prediction time, and the RMSE and R^2 fit parameters:

	Training		Prediction			
Model	CPU Time (ms)	Wall Time (ms)	CPU Time (ms)	Wall Time (ms)	RMSE	R^2
Linear Regression	31.2	2,710	0	4.6	0.72	0.60
scikit-learn CART	172	2,460	0	4.49	0.71	0.62
scikit-learn GradientBoosting Regressor	6,480	12,500	15.6	19.7	0.49	0.82
scikit-learn RandomForestRegressor	13,400	19,400	62.5	163	0.51	0.82
XGBoost	8,920	2,360	0	9.61	0.49	0.82

Table 5.3 – Table of results from testing each model

Comparing the R^2 and RMSE values among the models, the gradient-boosted and random forest regressors from scikit-learn, and XGBoost provides better predictions of the testing dataset than the simpler linear or classification and regression tree models. However, as you've measured, these ensemble models take longer to train. XGBoost takes roughly the same amount of training time as the other ensemble models while being more efficient when making a prediction.

Summary

In this chapter, you explored when to use ensemble methods versus linear regression and CART by building models with various algorithms using the housing value dataset. You compared XGBoost to linear regression, CART, gradient boosting, and random forest ensemble methods.

You learned when each of these models is a good choice and when to use deep learning. You ended this chapter by learning about the settings that control the learning process for the XGBoost algorithm.

Now that you have a broad understanding of XGBoost, you'll learn how to handle practical problems with data. In the next chapter, you will learn about methods for cleaning data, how to best handle imbalanced data when building a classifier, and how to deal with other data-related problems.

Data Cleaning, Imbalanced Data, and Other Data Problems

This chapter covers how to address common problems with real-life datasets. You will learn about data exploration and cleaning in more depth than was covered in *Chapters 2* and *4*. By the end of this chapter, you will know how to clean different types of data, whether the data is continuous numeric values, text categories, or date data. You will see what you should look for when exploring data and how to handle common data-cleaning problems. You will gain an understanding of how to manage unbalanced data for classification problems and how to work with transformed data.

In this chapter, you will learn about the following main topics:

- Real-life data is never clean
- What to look for when exploring data
- Data-cleaning methods
- Handling imbalanced data

Technical requirements

The code presented in this chapter consists of Python functions to perform common data-cleaning functions. The code is available on our GitHub repository: `https://github.com/PacktPublishing/XGBoost-for-Regression-Predictive-Modeling-and-Time-Series-Analysis`. You will need to install the software and Python packages in the following list to follow along with the chapter:

- Python 3.9 (a virtual environment is recommended)
- `pandas` 1.4.2
- Jupyter Notebook

- VS Code

Real-life data is never clean

So far, for the examples in *Chapters 2* and *4*, you have used two datasets from `scikit-learn`. These are both well-known and well-studied datasets. They may have been pre-cleaned. Datasets on Kaggle or in classes are often similarly clean datasets. Unlike these examples, data for real-life problems is never clean. You will find missing data, values that make no sense, inconsistencies in naming, and other problems. Typically, the majority of time spent on a data science project is spent cleaning and preparing the data for modeling. This includes time spent exploring the data, discussions with domain experts to understand the data, and time spent cleaning the data. As you learned in *Chapters 2* and *4*, spending time early in the process to visualize the data by building graphs will save time later on during the modeling process.

What to look for when exploring data

As mentioned in *Chapters 2* and *4*, we like to make graphs to explore data and understand what is contained in a dataset and what potential problems might exist. The tools and methods used to graph the data depend on the size of the dataset and what format the data is in. For example, if a smaller dataset is in Excel format or CSV format, it may be easier and faster to use Excel to explore the data. Larger datasets are more easily explored in Python with `pandas`, especially if they exceed the size limits for Excel to load them fully. Generally, you are checking for problems in the data such as missing values, values that don't make sense, misspellings in text data, and so on. You also want to look for relationships between the input parameters and between the input parameters and the parameter you want the model to predict. When you are comfortable with how the data looks and you have addressed problems with the data by cleaning the data, you can move on to building a model. *Table 6.1* shows how to check for problems and what to look for by the type of data:

Data Type	Check By	Look For
Continuous (numeric)	Plotting a distribution (histogram) Plotting x versus y scatterplots	Discontinuity Outliers The shape of the distribution – is it a normal distribution, skewed, bimodal, or multimodal? Consistency in the unit of measure

Ordinal (numeric, text) Values belong to ordered categories (for example, grades in education – A, B, C, D, F – or rankings in a survey – Excellent, Good, Fair, Poor – should be used in that order)	Plotting x versus y scatterplots If date data, plotting trend graphs	Discontinuity Outliers Consistency in spelling Categories are orthogonal Balanced categories – a similar number of rows in each category
Nominal (numeric, text) Values belong to categories, but the order is not important	Plotting distributions by category	Consistency in spelling Categories are orthogonal Balanced categories
Date	Plotting a distribution (histogram) Plotting trend graphs	Discontinuity Consistency in date format Consistency in time zone Seasonality
Currency	Plotting a distribution (histogram) Using a filter to preview the data in summarized form	Consistency in format Consistency in the unit of measure

Table 6.1 – Data problems to check for

This table summarizes what to look for when exploring data and common problems you may encounter that require data cleaning. In general, you are checking for inconsistencies in the data and missing data. You want to be sure all values in a column use a consistent unit of measure. For a classifier, you want to check that you have balanced data.

Beyond what is listed in the preceding table, for date data, you want to ensure you manage the time zone correctly. Some databases store datetime values in **Coordinated Universal Time (UTC)**, and some use the local time of the database, so make sure you understand how the data is stored. You may want to generate additional columns for shifts or workweek. If the database has a calendar table, use that table when you do any conversions.

If continuous data is **bimodal** or **multimodal**, meaning they have two peaks (bimodal) or multiple peaks (multimodal), do some investigation to see if separation by a categorical variable you have in the dataset explains the differences in means. For example, the iris data petal length parameter appears to be bimodal. Splitting this column by species yields individual normal distributions for the three categories, so the data is multimodal. To normalize this data, first split it by species, then do the normalization and recombine the normalized values into a single column. *Figure 6.1* illustrates this example:

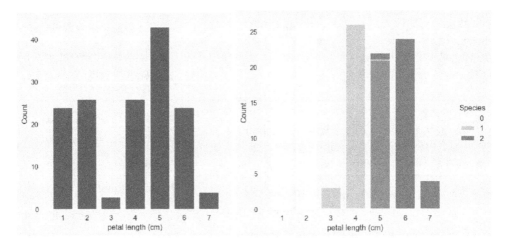

Figure 6.1 – Graph of the distribution of multimodal data using the Iris dataset as an example

In the graph on the left of *Figure 6.1*, you see a histogram of the combined measurements for petal length, which, while it appears to be bimodal, is actually multimodal. There are, in fact, three peaks, which when combined give the appearance of only two peaks. The graph on the right shows the same data separated by species, and the three peaks are now clear. You can normalize petal length for each species separately and then use the combined column. This example sits on the border between data cleaning and **feature engineering**. Creating additional columns based on existing data is called feature engineering and is handled in depth in *Chapter 7*.

In the next section, we will discuss data-cleaning methods you can use to address the problems uncovered when exploring the data.

Data-cleaning methods

Now that you have examined the dataset for problems, you need to address and correct the problems you've found. This is commonly the most time-consuming part of a project. When doing data cleaning, keep all the existing data and work in a new column when changing values (cleaning data). If dropping rows, create a new version of the data table. This allows you to go back to the original data easily if needed and gives you options when doing feature engineering, which you will learn about in *Chapter 7*.

> **Split your dataset before you clean the data to avoid inflating model accuracy**
>
> Before you do any data cleaning, split your data into training and test sets. This prevents leaking information from the training set into the testing set, which will inflate the results of accuracy tests. For example, if you calculate a mean of the dataset in full before you do the train/test split and have that data in a column in both datasets, you've provided the testing dataset with information from the training dataset. Do the *same* cleaning operations on *both* your training and test datasets.

The pandas and scikit-learn libraries have built-in functions that are helpful when doing data cleaning. The pandas library's `dataframe.replace` method allows you to modify values. For example, you can remove invalid data. Say, if a value of 9999 is used to indicate invalid data, you can call the `.replace` method and change 9999 to numpy's NaN value signifying the entry is "not a number" and should be ignored – `np.NaN`. The inverse, `dataframe.fillna`, allows you to set all missing values to a particular number. Depending on the dataset, you may use both options when doing data cleaning.

The scikit-learn library includes data-cleaning functions as well. It has a preprocessing package called `sklearn.preprocessing`. This package contains functions for the standardization of data, performing non-linear transformations, and normalizing data, among others.

We will address data cleaning for each data type (continuous, nominal, date) separately, starting with a table of common methods and providing code for Python functions to perform common cleaning tasks. Python functions are used so that you can reuse the cleaning code easily.

Let's create a test DataFrame to practice data-cleaning methods. You can use this DataFrame to verify that the functions work correctly:

1. First, make a pandas DataFrame with 5,250 rows and 3 columns containing random values using the following code:

    ```
    testdf = pd.DataFrame(np.random.randn(5250,3),
        columns=list("ABC"))
    ```

2. Then, add a `category` column:

    ```
    testdf.insert(len(testdf.columns), "category", "Category A")
    ```

3. Set the value for rows 3,000 to 4,499 to `"CategoryB"` and rows 4,500 to 5,250 to `"Cat C"`:

    ```
    testdf.loc[3000:4499,["category"]] = "CategoryB"
    testdf.loc[4500:5250,["category"]] = "Cat C"
    ```

    These names are deliberately mismatched so that you can practice correcting text data.

4. Finally, create a `Date` column and populate it with random data values using `np.random.choice` and `pd.daterange`:

```
testdf.insert(len(testdf.columns),"Date",np.random.choice(
    pd.date_range('2022-10-01', '2024-11-30'),len(testdf)))
```

This makes a DataFrame called `testdf` with five columns: A, B, C, `category`, and `Date`.

Cleaning continuous data

This table lists common data problems with continuous numeric data and the methods to use to clean data with those problems. In this section, you will create functions for removing outliers and for normalizing data:

Continuous Data Problem	Data-cleaning Method
Consistency in the unit of measure	Create a new column with the values converted to a consistent unit of measure
Outliers	Create a new column with filtered data – for example, after applying a 4-pseudo sigma filter
The shape of the distribution	Create a new column with the data normalized
Discontinuity	Impute values

Table 6.2 – Summary of common problems and corresponding data-cleaning methods

4-pseudo sigma filter for removing outliers from continuous data

In this section, we will create a Python function to perform filtering on a column of continuous data to remove outliers. It creates a new column with the filtered values using NumPy's `np.NaN` method to mark outliers. You can then create a new table and drop the rows with NaN to yield a cleaned dataset:

1. We'll start by creating a function to do a 4-pseudo sigma filter:

```
# Filter continuous data using a 4 pseudo sigma filter
# dataframe = pandas dataframe to modify
# parameter = name of column in dataframe to filter

def pseudosigmafilter(dataframe, parameter):
    mean = dataframe[parameter].mean()
    stdev = dataframe[parameter].std()
    lowerfiltervalue = mean - (4*stdev)
    upperfiltervalue = mean + (4*stdev)
```

```
# print(mean)
# print(stdev)
# print(lowerfiltervalue)
# print(upperfiltervalue)
```

For this, we first need to calculate the mean and the standard deviation for the column to filter. A 4-pseudo sigma filter is defined as the mean ± (4 x standard deviation). The DataFrame and column to filter are passed to the function as `dataframe` and `parameter`. We use `mean = dataframe[parameter].mean()` to calculate the mean and `stdev = dataframe[parameter].std()` to calculate the standard deviation. Then, we set `lowerfiltervalue` to `mean - (4*stdev)` and `upperfiltervalue` to `mean + (4*stdev)`.

2. Then, we will filter the data by creating a new column named `"filter_" + parameter` and setting it to NaN if the value of that row is outside the limits:

```
dataframe["filter_" + parameter] = np.where(
        ((dataframe[parameter] > lowerfiltervalue) & (
            dataframe[parameter] < upperfiltervalue)) ,
        dataframe[parameter], np.NaN)
    return dataframe
```

We create the new column with `dataframe["filter_" + parameter]`. We use NumPy's `where` function to create a conditional statement. If the value is within the limits `((dataframe[parameter] > lowerfiltervalue) & (dataframe[parameter] < upperfiltervalue))`, we set the value in the `filter` column to match the original value, `dataframe[parameter]`. If the value is outside the limits, set the `filter` column to `np.NaN`. End by returning the modified DataFrame.

This results in a new column being added to the DataFrame passed to the function. If a value in the column being filtered is outside the 4-pseudo sigma limits, the value in the new column is NaN; otherwise, it is the same as the original column value.

3. Finally, we test that the filter works by injecting a large value into a column and calling the `pseudosigmafilter` function. First, we change the value of the first row of the A column to 11 by setting `testdf.at[0,"A"] = 11`. Next, we call the filtering function:

```
testdf.at[0,"A"] = 11
pseudosigmafilter(testdf, "A")
```

Now there is a new column called `filter_A`, and the first row is set to NaN to filter the outlier value as shown here:

	A	B	C	category	Date	filter_A
0	11.000000	-0.467456	-0.644003	Category A	2023-10-23	NaN
1	0.317928	0.180459	-0.955891	Category A	2023-08-26	0.317928
2	1.717734	-1.461808	1.417787	Category A	2023-03-23	1.717734
3	-0.205581	0.976843	0.723133	Category A	2023-09-08	-0.205581
4	2.101230	0.369219	1.420681	Category A	2023-07-05	2.101230
...	...	...	...	...	...	...
5245	-0.497334	-0.955604	-1.729412	Cat C	2023-04-03	-0.497334
5246	-1.216212	-1.632568	0.207243	Cat C	2022-12-16	-1.216212
5247	0.244406	-0.546770	-0.006604	Cat C	2023-05-27	0.244406
5248	0.403877	-0.634075	1.316784	Cat C	2023-01-23	0.403877
5249	1.353647	-0.595278	-0.204126	Cat C	2023-03-01	1.353647

5250 rows × 6 columns

Figure 6.2 – Result of filtering testdf

In this section, you created a function that filters data for outliers by creating a column that tags values that fall outside the mean ± 4-pseudo sigma with NaN. Summary statistics calculated on the new `filter_A` column will automatically ignore outlier values. Next, you'll learn how to standardize a column of values.

Standardizing continuous data

Standardizing is a statistical process that puts values on the same scale, which allows you to compare values across different parameters. Standardizing data is important if you have a model that contains polynomial or interaction terms – for example, if you are modeling the impact on etch rate in a reactive ion etch operation in semiconductor manufacturing. In this process, there are non-linear interactions between the etch rate and parameters of the plasma used in the etch process. In these cases, without standardizing the input variables, you might produce misleading results.

Standardizing is achieved by subtracting the parameter mean from a given input parameter value and dividing it by the standard deviation. The good news is that there is a function built into scikit-learn to do this. Standardizing does not maintain the shape of the variables' distribution because it scales it by dividing by the standard deviation. The scikit-learn function allows you to turn off the scaling and just **center** the variable.

In this section, you will use the `StandardScaler` utility class in scikit-learn to standardize continuous data:

1. You'll need the `preprocessing` module from scikit-learn, so import that using `from sklearn import preprocessing`. The `StandardScaler` utility class works on continuous variables and will give an error if provided with categorical columns, so create a DataFrame that only contains columns A, B, and C – `continuous = testdf[["A", "B", "C"]]`:

    ```
    from sklearn import preprocessing
    continuous = testdf[["A", "B", "C"]]
    ```

    This makes a DataFrame called `continuous` that only has columns A, B, and C, as shown in *Figure 6.3*:

	A	B	C
0	11.000000	0.276126	-1.584829
1	-0.218692	0.644166	0.374023
2	-0.489941	-0.462914	-1.076924
3	0.225133	0.635604	0.366959
4	-0.070915	-0.903729	-2.236315

    Figure 6.3 – Result of continuous.head()

2. Next, you need to fit `StandardScaler` to the `continuous` columns. We can do this with the following code:

    ```
    standardized = preprocessing.StandardScaler().fit(continuous)
    ```

    This calculates the mean and standard deviation that will be used to center and scale the values. The result is a `StandardScaler` object called `standardized`.

3. Finally, apply the standardization to the dataset with the `continuous` columns:

    ```
    standardized = standardized.transform(continuous)
    print(standardized)
    ```

    This results in the following output:

    ```
    [[10.96358242 0.69523468 -1.71637082]
    [-1.37721364 -0.75714924 0.02145564]
    [ 0.38625059 0.09980872 1.39928846]
    ...
    ```

```
[ 0.4982133 -0.64749348 -2.17434114]
[-1.30593287 -1.03903145 -0.91478772]
[-0.899012 -0.89173605 1.27405133]]
```

This is a NumPy array that has been centered based on the mean of the data in the dataset and scaled by the standard deviation.

4. The pandas DataFrame you began with has been changed in the process to a NumPy array. You can convert it back to a pandas DataFrame with the following:

```
standardizeddf = pd.DataFrame(standardized, columns=[
    "A", "B", "C"] )
```

5. You can then concatenate it to your existing DataFrame using pd.concat like this:

```
standardizedtestdf = pd.concat([testdf, standardizeddf],axis=1)
```

This combines the standardized columns with the original testdf DataFrame and calls the result standardizedtestdf. *Figure 6.4* shows the result:

	A	B	C	category	Date	filter_A	A	B	C
0	11.000000	-0.760867	-0.644580	Category A	2023-04-16	NaN	10.935958	-0.745161	-0.641530
1	1.550786	-0.455346	-1.149559	Category A	2022-12-22	1.550786	1.542617	-0.438538	-1.138580
2	1.337273	1.480686	-0.712892	Category A	2023-09-29	1.337273	1.330366	1.504476	-0.708769
3	0.107457	-0.137325	0.136342	Category A	2023-09-24	0.107457	0.107822	-0.119370	0.127131
4	0.334353	1.169766	-1.943095	Category A	2022-11-25	0.334353	0.333377	1.192435	-1.919657

Figure 6.4 – Result of standardizedtestdf.head()

You have now used scikit-learn's preprocessing module to standardize the three continuous variables in the test DataFrame.

If you wish to only center the data, then use the with_std=False switch when creating the StandardScaler object like this:

```
centered = preprocessing.StandardScaler(
    with_std=False).fit(continuous)
```

Centering shifts the scale of the parameter by subtracting a constant from every value. This redefines the zero point for that parameter. When using StandardScaler before modeling data, you can build it into a data processing pipeline to easily apply the same transformations to your train and test datasets.

Now that you have standardized continuous data, the next section will cover normalizing continuous data.

Normalizing continuous data

In this section, you will create a Python function that normalizes a column of continuous data. It creates a new column with the normalized values using the `normalize()` function from scikit-learn. This can help reduce model overfitting. The default norm for `normalize()` is the Euclidean norm or L2. You can choose other norms as needed, by using `norm` = and choosing `max`, `l1`, or `l2`. Let's begin:

1. You'll need the `preprocessing` module from scikit-learn, so import that to start:

   ```
   from sklearn import preprocessing
   ```

2. Create a function called `normalizecolumn` and have it accept a pandas DataFrame and the column to normalize as inputs:

   ```
   def normalizecolumn(dataframe, parameter):
   ```

3. The normalize function works on array-type objects, so put the column to normalize into a NumPy array:

   ```
   colarray = np.array(dataframe[parameter])
   ```

4. Then, call the normalize function on the array and turn the result into a list to get it ready to add to the DataFrame:

   ```
   normalizedarray = preprocessing.normalize(
       [colarray]).tolist()
   ```

5. This makes a row of normalized values. Convert the row to a column using `np.swapaxes`:

   ```
   normalizedarray = np.swapaxes(normalizedarray,0,1)
   ```

6. Finally, add the result to the DataFrame as a new column,:

   ```
   dataframe["normalized_" + parameter] = normalizedarray
   return dataframe
   ```

   The result of calling the `normalizecolumn` function is a new column in the DataFrame with the values of the column you passed to have a unit norm.

7. Test that the function works by calling it to normalize the values in column B:

   ```
   normalizecolumn(testdf, "B")
   ```

This results in the output shown in *Figure 6.5*:

	A	B	C	category	Date	filter_A	normalized_B
0	11.000000	-0.467456	-0.644003	Category A	2023-10-23	NaN	-0.006447
1	0.317928	0.180459	-0.955891	Category A	2023-08-26	0.317928	0.002489
2	1.717734	-1.461808	1.417787	Category A	2023-03-23	1.717734	-0.020160
3	-0.205581	0.976843	0.723133	Category A	2023-09-08	-0.205581	0.013472
4	2.101230	0.369219	1.420681	Category A	2023-07-05	2.101230	0.005092
...	...	...	...	...	...	...	...
5245	-0.497334	-0.955604	-1.729412	Cat C	2023-04-03	-0.497334	-0.013179
5246	-1.216212	-1.632568	0.207243	Cat C	2022-12-16	-1.216212	-0.022515
5247	0.244406	-0.546770	-0.006604	Cat C	2023-05-27	0.244406	-0.007541
5248	0.403877	-0.634075	1.316784	Cat C	2023-01-23	0.403877	-0.008745
5249	1.353647	-0.595278	-0.204126	Cat C	2023-03-01	1.353647	-0.008210

5250 rows × 7 columns

Figure 6.5 – Result of calling normalizecolumn(testdf, "B")

In this section, you learned about common problems in numeric data that require data cleaning and methods to handle those problems. You created functions to filter outliers, standardize data, and normalize data.

Next, you'll create a function to clean up spelling in a text column. This is helpful when you want to use the text as labels. The example here is pretty straightforward. For more complex datasets, we often use programs such as JMP from SAS Institute to clean up text columns. JMP has a feature called Column Recode that makes this type of data cleaning easy and interactive.

For many models, you will want to encode the text as numeric data. You will learn more about methods for handling categorical data in *Chapter 8*. Before we get to cleaning text data, let's first go over the types of problems you may encounter and how to address them. Then, we'll clean up text data that has spelling inconsistencies.

Cleaning ordinal and nominal data

This table lists common data problems with text data, ordinal and nominal data, and the methods to use to clean data with those problems. In this section, you will create a function to correct spelling within a column:

Ordinal and Nominal Data Problem	Data-cleaning Method
Discontinuity	Decide how to deal with the lack of data for a missing category – either collect more data, ignore missing values, or impute values
Outliers	Filter the categories, ignore outliers
Consistency in spelling	Create a new column with corrected spellings
Categories are orthogonal (no confusion or overlap as to how a sample should be categorized)	Create a new column with re-categorized data that is clearly differentiated
Balanced categories	See the Handing imbalanced data section

Table 6.3 – Summary of common problems with ordinal and nominal data and cleaning methods

Frequently ordinal and nominal data is in the form of text strings. Dealing with data cleaning of text in a dataset can be complex. Often, nominal data is the labels for a dataset and is important for modeling. This data is frequently entered by people who may not remember exactly what they called something the last time they labeled data, so you will see differences in capitalization, spelling errors, differences in where there are spaces between words, and other subtle differences. Unfortunately, a computer cannot tell that "State St," "State Street," and "state street" are all the same thing. Without data cleaning, your model will treat all these forms as different categories. There are many **natural language processing (NLP)** techniques for text data cleaning, which we will not cover here. The best way to deal with this problem, in our experience, is to automate data collection as much as possible so that there are fewer opportunities for variation.

In the next section, you will handle a very common problem, which is variations in spelling in a column.

Correcting spelling in a column

Here, a function will create a new column with corrected spellings. This function will be tailored to the example dataset. Making a more universal function gets into parsing and other topics in NLP and is beyond the scope of this book.

The `category` column in the `testdf` DataFrame has three values: `"Category A"`, `"CategoryB"`, and `"Cat C"`. This function will standardize the format to be `"Category"` followed by a space and then the letter A, B, or C depending on what was in the original column. This will give you some practice with manipulating text. You can then expand on these concepts when cleaning text data. we find that it sometimes takes less effort to fix data using a program such as JMP or Excel than to figure out how to make Python do it. Depending on how large your data is and how frequently the cleaning task will need to be done, consider using interactive methods rather than spending a lot of time figuring out how to clean the data. Let's begin correcting the spelling:

1. Create a function called `cleancategory` that takes `dataframe` and `parameter` as input: `def cleancategory(dataframe, parameter):`. Since this function only works for text and is tailored to the `category` column in particular, check that the parameter passed to the function is the `category` column and don't do anything if this is not the case: `if parameter == "category":`. You know by looking that the letter designating the category is at the end of the string, so use `dataframe[parameter].str[-1]` to get that letter and combine it with `"Category"`. Put this formatted string into a new column with `dataframe["clean_" + parameter] = "Category " + dataframe[parameter].str[-1]`.

2. Test that the function works by calling it to clean the `category` column.

```
cleancategory(testdf, "category" )
```

This results in a new column called `clean_category` being added to the table with the reformatted contents. *Figure 6.6* shows the output:

	A	B	C	category	Date	filter_A	normalized_B	clean_category
0	11.000000	-0.467456	-0.644003	Category A	2023-10-23	NaN	-0.006447	Category A
1	0.317928	0.180459	-0.955891	Category A	2023-08-26	0.317928	0.002489	Category A
2	1.717734	-1.461808	1.417787	Category A	2023-03-23	1.717734	-0.020160	Category A
3	-0.205581	0.976843	0.723133	Category A	2023-09-08	-0.205581	0.013472	Category A
4	2.101230	0.369219	1.420681	Category A	2023-07-05	2.101230	0.005092	Category A
...	...	...	...	...	...	...	...	...
5245	-0.497334	-0.955604	-1.729412	Cat C	2023-04-03	-0.497334	-0.013179	Category C
5246	-1.216212	-1.632568	0.207243	Cat C	2022-12-16	-1.216212	-0.022515	Category C
5247	0.244406	-0.546770	-0.006604	Cat C	2023-05-27	0.244406	-0.007541	Category C
5248	0.403877	-0.634075	1.316784	Cat C	2023-01-23	0.403877	-0.008745	Category C
5249	1.353647	-0.595278	-0.204126	Cat C	2023-03-01	1.353647	-0.008210	Category C

5250 rows × 8 columns

Figure 6.6 – Result of calling cleancategory(testdf, "category")

Here, we learned about the common types of data-cleaning problems for ordinal and nominal data. You created a function that standardizes the spelling of text in a column by aligning the naming used for categories. Next, you'll learn about problems that can occur with date data and create a function to reformat date data.

Reformatting date data

This table lists common data problems with date data and the methods to use to clean data with those problems. In this section, you will create a function to reformat the dates in a column:

Date Data Problem	Data-cleaning Method
Discontinuity	Decide how to deal with the lack of data – either collect more data, ignore missing values, or impute values
Consistency in date format	Create a new column with the date data in a standard format
Consistency in time zone	Create a new column and convert the times to a uniform time zone
Seasonality	Create a new column for the parameter experiencing variations with season and standardize it

Table 6.4 – Summary of common problems with date data and cleaning methods

When converting to a uniform time zone, you can use UTC or convert to the time zone of your choice that makes sense for your use case. we recommend including the time zone as part of the name of the column if it is important.

Next, you will write a function to convert date data from one format to another. Specifically, we will convert a date formatted as year, month, day to year-month in words – date – and then month/day/year. This will give you examples of date format conversions in Python and pandas, which you can build on to convert to other date and time formats as needed:

1. The Date column in testdf is currently in the format of year, month, and day. This is sometimes written as yyyy-mm-dd. Change that to a format that uses a three-letter abbreviation for the month, sometimes written as dd-mon-yyyy. This format is useful in removing confusion as to whether the day or the month is being listed first. You will do this by creating a new column called Date_new using the pandas dt.strftime method. This method uses standard Python string formats, which you can read more about at https://docs.python.org/3/library/datetime.html#strftime-and-strptime-behavior. %d gives the day of the month with a zero in front of it if less than 10. %b gives the month as a three-letter abbreviation – for example, Sep for September. %Y gives the year:

    ```
    testdf["Date_new"].dt.strftime("%d-%b-%Y")
    ```

This results in a new column in the DataFrame that contains the date converted to a string in the format dd-mon-yyyy:

	A	B	C	category	Date	filter_A	Date_new
0	11.000000	-0.760867	-0.644580	Category A	2023-04-16	NaN	16-Apr-2023
1	1.550786	-0.455346	-1.149559	Category A	2022-12-22	1.550786	22-Dec-2022
2	1.337273	1.480686	-0.712892	Category A	2023-09-29	1.337273	29-Sep-2023
3	0.107457	-0.137325	0.136342	Category A	2023-09-24	0.107457	24-Sep-2023
4	0.334353	1.169766	-1.943095	Category A	2022-11-25	0.334353	25-Nov-2022

Figure 6.7 – Reformatting date to dd-mon-yyyy

This example serves to illustrate the benefit of using a new column when changing formats. The formatting has changed the data type of the column, but since you did it in a new column, you still have the original date data in the original data type in the Date column.

2. Next, let's create another new column called Date_US that formats the date in the mm/dd/yyyy format common in the US. Use the dt.strftime method again, this time with %m for the two-digit month, %d for the day of the month, and %Y for the year. Notice that you control the separator between the values as well. In *step 1*, you used dashes, and in this step, you use a slash:

```
testdf["Date_US"] = testdf["Date"].dt.strftime("%m/%d/%Y")
```

This will add a new Date_US column, as shown here:

	A	B	C	category	Date	filter_A	Date_new	Date_US
0	11.000000	-0.760867	-0.644580	Category A	2023-04-16	NaN	16-Apr-2023	04/16/2023
1	1.550786	-0.455346	-1.149559	Category A	2022-12-22	1.550786	22-Dec-2022	12/22/2022
2	1.337273	1.480686	-0.712892	Category A	2023-09-29	1.337273	29-Sep-2023	09/29/2023
3	0.107457	-0.137325	0.136342	Category A	2023-09-24	0.107457	24-Sep-2023	09/24/2023
4	0.334353	1.169766	-1.943095	Category A	2022-11-25	0.334353	25-Nov-2022	11/25/2022

Figure 6.8 – Reformatting date to mm/dd/yyyy

The testdf DataFrame now has a column called Date_US with the value of the Date column as a string in the mm/dd/yyyy format.

In this section, you learned about common types of problems with date data and used the pandas dt.strftime method to convert from one date format to another.

You now have an understanding of the various methods for cleaning different types of data. Next, you will learn about what to do when you are building a classifier model and you don't have the same number of examples in each category.

Handling imbalanced data

Imbalanced data is a challenge for classification problems. Because **machine learning** (**ML**) models such as XGBoost learn from historical data, if you don't have examples, your model cannot learn the pattern. If your data only has 3 samples for a particular category, then the model can't learn the pattern that predicts a member of that category as effectively as if it had 3,000 samples. Additionally, if you have two categories (binary classifier) and one category has many more members than the other in the training data, you essentially train the model to predict just that category. Think about it this way – imagine you have been asked to predict the color of a ball pulled out of a bag. Every time you observe a ball being pulled out, it has been red. What color would you guess next? Red, of course. You would be surprised to see a ball of a different color, but not shocked – especially if told in advance that there are two possibilities. Similarly, if you have a dataset that has many examples of category *A* and only a handful of category *B*, then your classifier will have been taught to always predict that the answer is *A*.

You can correct imbalanced data in two ways: sampling the dataset, and enhancing the dataset. We will address each in turn next.

Correcting imbalanced data by sampling the dataset

If you have a large number of observations (rows) in your data, you can adjust imbalanced data by randomly selecting data by category and matching the number of observations in the smallest category. For example, say you have a dataset with three categories, *A*, *B*, and *C*. *A* has 3000 rows of data, *B* has 1500 rows, and *C* has 750 rows. To balance the data across the three classes, you would randomly select 750 rows of category *A* and 750 rows of category *B* and keep all of category *C*. Your resulting dataset would have a total of 2250 rows.

To do this, use `pandas` to select a random sample by category. This code creates a function called `subsamplecategory` that creates a new DataFrame containing a random sample of `nsample` rows of each value in the category:

```
def subsamplecategory(olddataframe, category, nsamples):
    newdataframe = olddataframe.groupby(category).apply(lambda s:
        s.sample(nsamples))
    return newdataframe
```

Now, let's see how can we enhance data by imputing values.

Enhancing data for a class by imputing values

A simple method for enhancing data for a category with few samples (a minority class) is to oversample the category, meaning repeat the existing data points multiple times. This ensures the model does not ignore the minority category, which is what happened in our example with pulling out a ball from the bag but does not add any new information for the model.

Another option is to oversample the category by creating synthetic data based on imputing values from the existing data. One method to do this is called **Synthetic Minority Oversampling TEchnique (SMOTE)**, which is described in a 2002 paper by N.V. Chawla et al., published in the *Journal of Artificial Intelligence Research, SMOTE: Synthetic Minority Over-sampling Technique*. The paper shows that a combination of under-sampling of the majority category and oversampling of the minority category can achieve better classifier performance. This algorithm has been implemented in the `imbalanced-learn` Python library. This article, `https://machinelearningmastery.com/smote-oversampling-for-imbalanced-classification/`, has more information about SMOTE and the `imbalanced-learn` library.

> **Working with transformed data**
>
> If you transformed data to create your model – for example, if you standardized or normalized the inputs – then when using the model to make predictions (inference), you will need to perform the same transformation on the input variables and then reverse any transformation on the output variable. Otherwise, you will not get the correct results because the output will be centered and scaled.

Summary

In this chapter, you learned how to clean different types of data, whether the data is continuous numeric values, text categories, or date data. You learned what you should look for when exploring data and how to handle common types of data-cleaning problems. You created Python functions to handle common data-cleaning tasks. You gained an understanding of how to manage imbalanced data for classification problems and how to work with transformed data. All of these will help you make sure your model's accuracy is not hampered by data problems.

In the next chapter, you will learn more about feature engineering, which is creating new input columns from existing data, and feature selection, where you reduce the number of input columns to those with the most effect on the model. In that chapter, we will be using a housing price dataset from Kaggle.com, which can be found at `https://www.kaggle.com/competitions/house-prices-advanced-regression-techniques/`. In preparation for that chapter, you can review the **exploratory data analysis (EDA)** Jupyter notebook found on our repository: `ch7/exploratory_data_analysis.ipynb`. The notebook uses the techniques provided in this chapter on the dataset that you will use in *Chapter 7*.

7

Feature Engineering

This chapter covers various feature engineering topics while providing source code examples so that you can gain proficiency with these techniques and apply them to real-world predictive modeling scenarios. We'll be using the house prices dataset as a practical illustration. By the end of this chapter, you'll acquire not only theoretical knowledge but also practical skills to navigate the world of data processing.

In this chapter, we'll cover the following topics:

- A review of exploratory data analysis

- Performing feature engineering on the house prices dataset

- Common feature engineering techniques for numerical features

- Common feature engineering techniques for temporal features

- Common feature engineering techniques for categorical features

> **Split your dataset before performing feature engineering to avoid inflating model accuracy**
>
> As you learned in *Chapter 6*, it's critical to divide the dataset into distinct training and testing subsets before performing any feature engineering. This practice ensures the integrity of model evaluation and guards against overfitting, thereby enhancing the reliability and generalizability of predictive models.

Technical requirements

This chapter serves as a hands-on guide, offering practical insights and techniques for harnessing the predictive power of numerical and temporal variables. Through a blend of theoretical discourse and code demonstrations, you'll gain proficiency in leveraging Python libraries to navigate the intricacies of data preprocessing. The code presented in this chapter is available in this book's GitHub repository: `https://github.com/PacktPublishing/XGBoost-for-Regression-Predictive-Modeling-and-Time-Series-Analysis`.

This chapter relies on the following Python libraries, all of which should be installed before you proceed:

- pandas 1.4.2
- NumPy
- Matplotlib
- Seaborn
- SciPy

A review of exploratory data analysis

This chapter uses the house prices dataset. This dataset is a valuable resource for understanding the dynamics of the real estate market. In this section, you'll review **exploratory data analysis (EDA)**, aiming to unravel the underlying insights within this dataset. To perform this analysis, you'll use essential Python libraries such as pandas, NumPy, Matplotlib, and Seaborn.

Dataset and data description

You can access the house prices dataset via Kaggle: `https://www.kaggle.com/competitions/ house-prices-advanced-regression-techniques/data`. This is a different dataset than the California house prices dataset you used in *Chapter 4*. To understand the variables within the dataset, you can refer to the accompanying data description: `https://www.kaggle. com/competitions/house-prices-advanced-regression-techniques/ data?select=data_description.txt`. You'll be using this dataset in subsequent chapters as well.

> **Citation for the house prices dataset**
> Dean De Cock (2011) Ames, *Iowa: Alternative to the Boston Housing Data as an End of Semester Regression Project, Journal of Statistics Education*, 19:3, DOI: 10.1080/10691898.2011.11889627.

Performing EDA

Before you begin feature engineering, you need to fully understand the dataset. You learned about this type of analysis in *Chapter 6*, so we won't revisit this here. We've created a separate Jupyter notebook that you can refer to and explore the house prices dataset. In this Jupyter notebook, you'll find the following analyses:

- Analyzing the target variable
- Understanding variable data types
- Checking for missing data

- Analyzing numerical variables
- Analyzing categorical variables

This is the same type of analysis that you did in *Chapter 6* but applied to this house prices dataset. For that reason, we won't cover it in detail here. The Jupyter notebook is available on GitHub: `https://github.com/PacktPublishing/XGBoost-for-Regression-Predictive-Modeling-and-Time-Series-Analysis/blob/partha1/ch7/exploratory_data_analysis.ipynb`. These analyses will help you understand what's included in the dataset and where you may have the opportunity to enhance the data through feature engineering. Let's look at each analysis in turn.

Analyzing the target variable

To begin, you'll want to examine the target variable, which represents housing prices. In this part of the analysis, you'll be using descriptive statistics and visualizations to gain insights into the distribution and variability of housing prices, setting the stage for predictive modeling later on.

Understanding variable data types

Next, you'll want to categorize the variables into categorical and numerical types. By delineating variables based on their data types, you set the groundwork for subsequent analysis strategies and will be able to extract insights from the data through feature engineering.

Missing data analysis

Once you've separated the variables into categories by type, you can address missing data. Missing data is handled differently for categorical variables and numeric variables, which is why you perform data type analysis beforehand. Missing data causes problems in data analysis and modeling. Through this analysis, you can quantify the extent of missing data within the dataset and explore strategies for handling it, ensuring the robustness and reliability of your models.

Analyzing numerical variables

Numerical variables, ranging from discrete to continuous, offer rich insights into the dynamics of the housing market. By examining their distributions and performing transformations, you can uncover patterns and trends that underpin housing price dynamics.

Analyzing categorical variables

Categorical variables, with their diverse attributes and levels, warrant special examination. Through cardinality analysis, identifying rare labels, and implementing special mappings, you can refine categorical variables to enhance their predictive utility in modeling endeavors. You'll do some initial feature engineering work with categorical variables in this chapter in preparation for *Chapter 8*, which discusses this type of variable in detail.

In conclusion, EDA serves as a foundational step in uncovering insights and patterns within complex datasets such as the house prices dataset. By leveraging the insights that have been gleaned from this analysis, you can make informed decisions about the new features you might wish to engineer and derive actionable insights from the data. A detailed exploration can be found in the associated Jupyter notebook on GitHub: `https://github.com/PacktPublishing/XGBoost-for-Regression-Predictive-Modeling-and-Time-Series-Analysis/blob/main/ch7/exploratory_data_analysis.ipynb`.

Now that you've gained a deeper understanding of the house prices dataset through EDA, you're ready to do some feature engineering.

Performing feature engineering on the house prices dataset

Feature engineering is a key stage in the data preprocessing pipeline. It transforms raw data into informative features that drive predictive modeling. It's a process that involves creating additional input columns that are to be used when you're developing a model that improves the model's performance. For example, you may transform numeric values so that they match a normal distribution, or group values into buckets to be able to convert from continuous values into categories. Let's say that you're looking at ages. In this case, you can bucket the values into age ranges of 0-10, 11-20, 21-30, and so on.

In this section, you'll learn how to perform feature engineering by using the house prices dataset. The code for this chapter has been provided in a Jupyter notebook that can be found in this book's GitHub repository. It provides a detailed exploration of feature engineering techniques: `https://github.com/PacktPublishing/XGBoost-for-Regression-Predictive-Modeling-and-Time-Series-Analysis/blob/main/ch7/feature_engineering.ipynb`.

In this section, you'll practice performing feature engineering using a real-world example of the house prices dataset. This includes the following aspects:

- Handling missing values for categorical variables
- Handling missing values for numeric variables
- Handling temporal variables
- Performing transformations on numeric variables

Chapter 8, covers handling missing data for categorical variables, as well as encoding techniques.

> **Guiding principle – train-test separation**
>
> The guiding principle behind separating the dataset into training and testing sets is rooted in the notion of unbiased model evaluation. During feature engineering, certain techniques may learn parameters from the data, which means you must isolate the training set so that it serves as an exclusive source of information for model parameter estimation. Failure to adhere to this principle may compromise the validity and reliability of model evaluations, leading to inflated performance metrics and erroneous conclusions.

Before you can start performing feature engineering, there's one key step you must take, which is to separate the dataset into a training set and a testing set. You'll do that in the next section.

Separating the dataset into training and testing sets

Before we delve into various feature engineering techniques, it's key to split the data into distinct training and testing subsets. This initial step safeguards against information leakage and ensures the integrity of model evaluation. By segregating the dataset upfront, you can prevent information from being transferred inadvertently from the training set to the testing set, thereby preserving the generalizability of your predictive models. Let's get started:

1. Begin by importing the libraries you'll use to ingest the data and perform feature engineering:

    ```
    import pandas as pd
    import numpy as np
    import matplotlib.pyplot as plt
    import scipy.stats as stats
    from sklearn.model_selection import train_test_split
    from sklearn.preprocessing import MinMaxScaler
    import joblib
    ```

2. Next, to be able to see all the columns in the dataset with ease, change the `display.max_columns` option in `pandas` to None:

    ```
    pd.pandas.set_option("display.max_columns", None)
    ```

3. Load the dataset you saved from Kaggle using `read_csv`. To check that the dataset was read correctly, print `data.shape` and use `data.head()` to look at the first few rows:

    ```
    data = pd.read_csv("house_pricing.csv")
    print(data.shape)
    data.head()
    ```

This results in the following output:

```
(1460, 81)
```

	Id	MSSubClass	MSZoning	LotFrontage	LotArea	Street	Alley	LotShape	LandContour	Utilities	LotConfig	LandSlope	Neighborhood
0	1	60	RL	65.0	8450	Pave	NaN	Reg	Lvl	AllPub	Inside	Gtl	CollgCr
1	2	20	RL	80.0	9600	Pave	NaN	Reg	Lvl	AllPub	FR2	Gtl	Veenker
2	3	60	RL	68.0	11250	Pave	NaN	IR1	Lvl	AllPub	Inside	Gtl	CollgCr
3	4	70	RL	60.0	9550	Pave	NaN	IR1	Lvl	AllPub	Corner	Gtl	Crawfor
4	5	60	RL	84.0	14260	Pave	NaN	IR1	Lvl	AllPub	FR2	Gtl	NoRidge

Figure 7.1 – Result of data.head()

4. You can use `train_test_split` to separate the dataset into train and test groups, just as you've done in previous chapters. You can also take this as an opportunity to clean up the dataset a bit. For example, you don't need the `Id` column as it's just an index, which you get when you put a CSV file into a pandas DataFrame. You can drop it and `SalePrice`, which is our target variable, from the X datasets with `data.drop (['Id', 'SalePrice'], axis=1)`. You can select 10% of the data to be randomly selected as the test set with `test_size=0.1`, and use `random_state = 0` as the random seed:

```
X_train, X_test, y_train, y_test = train_test_split(
    data.drop(["Id", "SalePrice"], axis=1),
    data["SalePrice"], test_size=0.1, random_state=0)
```

5. Now, you should have a training set with 1,314 rows and 79 columns and a test set with 146 rows and 79 columns. You can check this with `X_train.shape, X_test.shape`:

```
X_train.shape, X_test.shape
```

This results in the following output:

```
((1314, 79), (146, 79))
```

With that, you've loaded the dataset into a DataFrame and split it in preparation for modeling. Now, you can start performing feature engineering. You'll start by managing missing values.

Handling missing values for categorical variables

Our dataset contains instances of missing values within certain categorical variables. In this section, you'll learn about some strategies you can implement to address these missing values, along with why you should use the various approaches. The first step for dealing with missing values is to understand which variables have this problem. You can follow the processes that were outlined in *Chapter 6* to identify the variables with problems.

But before we get into that, there's one change we need to make to the dataset: we must add the MSSubClass variable to the list of categorical variables. You can learn about this variable by reading the data description in the data_dscription.txt file. This is available alongside the dataset on the Kaggle website: https://www.kaggle.com/competitions/house-prices-advanced-regression-techniques/data?select=data_description.txt. The MSSubClass variable is a number that encodes the type of house involved in the sale. This table lists the codes that are used and their meaning:

MSSubClass Code	Description
20	1-STORY 1946 & NEWER ALL STYLES
30	1-STORY 1945 & OLDER
40	1-STORY W/FINISHED ATTIC ALL AGES
45	1-1/2 STORY - UNFINISHED ALL AGES
50	1-1/2 STORY FINISHED ALL AGES
60	2-STORY 1946 & NEWER
70	2-STORY 1945 & OLDER
75	2-1/2 STORY ALL AGES
80	SPLIT OR MULTI-LEVEL
85	SPLIT FOYER
90	DUPLEX - ALL STYLES AND AGES
120	1-STORY PUD (Planned Unit Development) - 1946 & NEWER
150	1-1/2 STORY PUD - ALL AGES
160	2-STORY PUD - 1946 & NEWER
180	PUD - MULTILEVEL - INCL SPLIT LEV/FOYER
190	2 FAMILY CONVERSION - ALL STYLES AND AGES

Table 7.1 – MSSubClass codes and their meanings

The MSSubClass variable is an example of why it's helpful to consult with documentation about the dataset or with data domain experts when performing featuring engineering. If you hadn't read about the meaning behind these codes, you may have come to an incorrect conclusion regarding the value of the MSSubClass variable in your model. The following bit of code adds MSSubClass to a list of categorical variables and casts all categorical variables as type objects:

1. Build a list of all the categorical variables by looking for the variables that aren't numbers. You can check whether a variable isn't a number by looking for Python objects or dtype == 'O':

```
vars_cat = [var for var in data.columns if data[
    var].dtype == "O"]
```

2. Next, add 'MSSubClass' to the list:

```
vars_cat = vars_cat + ["MSSubClass"]
```

3. Then, you can cast all the categorical variables as the "O" object type and get a count of the number of categorical variables in the dataset. Since the list of columns is the same for both the train and test splits, we can use `vars_cat` for both datasets:

```
X_train[vars_cat] = X_train[vars_cat].astype("O")
X_test[vars_cat] = X_test[vars_cat].astype("O")
len(vars_cat)
```

This produces the following result:

```
44
```

Now, you have a list that contains all the names of the categorical columns. In the next section, you'll use this list to check the categorical columns for missing values.

Identifying variables with missing values

If there aren't a lot of values in a column, then we say that it has a high proportion of missing values. You can check for missing values in a column by checking each value in the column with `isnull()`. If it's empty, `isnull()` will be `True`. Then, you can sum up the `isnull()` values to get a numeric value of how many rows in a column have missing data. This is explored in more detail in the following example from the `exploratory_data_analysis.ipynb` Jupyter notebook, which can be found in this book's GitHub repository:

1. Build a list of missing variables that are of both the categorical and numeric type by looking for `isnull()`:

```
cat_vars_with_na = [
    var for var in vars_cat if X_train[
        var].isnull().sum() > 0]
```

2. Next, sort the variables by the proportion of missing data:

```
X_train[cat_vars_with_na].isnull().mean().sort_values(
    ascending=False)
```

This results in the following output:

```
PoolQC          0.995434
MiscFeature     0.961187
Alley           0.938356
Fence           0.814307
FireplaceQu     0.472603
GarageType      0.056317
```

```
GarageFinish      0.056317
GarageQual        0.056317
GarageCond        0.056317
BsmtExposure      0.025114
BsmtFinType2      0.025114
BsmtQual          0.024353
BsmtCond          0.024353
BsmtFinType1      0.024353
MasVnrType        0.004566
Electrical        0.000761
dtype: float64
```

This listing specifies each variable and how much of the column contains missing data. The data has been sorted, so the columns with the most missing values are given first. Here, PoolQC, MiscFeature, and Alley are the top three. Converting the proportion into a percentage, PoolQC has more than 99% missing values (0.995434), while MiscFeature has more than 96% missing (0.961187). These variables having missing values makes sense since not all houses will have a pool, need to capture information about miscellaneous features, or have an alley.

Now that you know how many categorical variables have missing values, let's implement a threshold-based strategy to address them.

Correcting missing categorical values with threshold selection and imputation

Missing values can be handled in several ways. For instance, you can drop the column from the training dataset using the pandas .drop method (for example, dataframe.drop(columns=['column', 'column'])), which means you ignore the column when building your model. Another way to handle missing data is to impute it, which involves deriving what the value should be based on what else is known about the row in question. For example, if you know a house has two stories, you can assume it has stairs. There are limits to this, of course. For example, if you know a house has three bedrooms, you can't say for certain that it has three bathrooms.

To guide the missing value imputation strategy, you can set a threshold criterion, flagging variables with more than 10% missing for special treatment. You can change this threshold to meet the needs of your dataset, based on what you know about the data. Let's take a look:

1. You can put the variables that meet the criteria into a list called for_missing_string and then replace the missing values with 'Missing' to preserve the integrity of the dataset:

    ```
    for_missing_string = [var for var in cat_vars_with_na if
        X_train[var].isnull().mean() > 0.1]
    ```

2. You can print for_missing_string to see the result:

    ```
    print(for_missing_string)
    ```

This gives us the following output:

```
['Alley', 'FireplaceQu', 'PoolQC', 'Fence', 'MiscFeature']
```

Now, you can replace the missing values with `'Missing'` for the variables in `for_missing_string` by using the `fillna` method. Ensure you do this for both `X_train` and `X_test`:

```
X_train[for_missing_string] = X_train[
    for_missing_string].fillna('Missing')
X_test[for_missing_string] = X_test[
    for_missing_string].fillna('Missing')
```

3. Missing values in variables below the threshold will be replaced by the most frequent value in the column. Put them in a list called `for_frequent_category`:

```
for_frequent_category = [
    var for var in cat_vars_with_na if X_train[
    var].isnull().mean() < 0.1]
```

To replace the missing values with the most frequent value in the column, you must find the most common value using the **mode** of the training dataset, and then use this value to replace the missing values in both the training and the test dataset. The reason you use the mode from the **training dataset** in both cases is that when you deploy your model and are using it for inference, you will often only have one row of data to use as input. If a value is missing in that instance, you only have your training dataset to use to calculate a replacement value. Using the same method during model development that you use during inference will ensure the results you get during training and testing will carry forward to model deployment. We'll do this next.

4. To calculate the mode for each column in the `for_frequent_category` list, you can use the `.mode()` method inside a `for` loop, and then take the first instance (if there are multiples) by selecting `.mode()[0]`. Lastly, use `X_train[var].fillna(mode, inplace=True)` and `X_test(var).fillna(mode, inplace=True)` to replace the missing values:

```
for var in for_frequent_category:
    mode = X_train[var].mode()[0]
    print(var, mode)
    X_train[var].fillna(mode, inplace=True)
    X_test[var].fillna(mode, inplace=True)
```

This prints out the following:

```
MasVnrType None
BsmtQual TA
BsmtCond TA
BsmtExposure No
BsmtFinType1 Unf
BsmtFinType2 Unf
Electrical SBrkr
```

```
GarageType Attchd
GarageFinish Unf
GarageQual TA
GarageCond TA
```

5. Next, you can check that all the missing and null values have been filled in by calculating the sum of the isnull() method for X_train[cat_vars_with_na]. If the sum is zero, all the missing values have been filled in:

    ```
    X_train[cat_vars_with_na].isnull().sum()
    ```

    The result is all zeros, so it worked on X_train:

    ```
    Alley           0
    MasVnrType      0
    BsmtQual        0
    BsmtCond        0
    BsmtExposure    0
    BsmtFinType1    0
    BsmtFinType2    0
    Electrical      0
    FireplaceQu     0
    GarageType      0
    GarageFinish    0
    GarageQual      0
    GarageCond      0
    PoolQC          0
    Fence           0
    MiscFeature     0
    dtype: int64
    ```

6. Now, you can repeat the test for X_test[cat_vars_with_na]:

    ```
    X_test[cat_vars_with_na].isnull().sum()
    ```

    The result is also all zeros for X_test:

    ```
    Alley           0
    MasVnrType      0
    BsmtQual        0
    BsmtCond        0
    BsmtExposure    0
    BsmtFinType1    0
    BsmtFinType2    0
    Electrical      0
    FireplaceQu     0
    GarageType      0
    ```

```
GarageFinish    0
GarageQual      0
GarageCond      0
PoolQC          0
Fence           0
MiscFeature     0
dtype: int64
```

Now that you've dealt with the missing values for the categorical variables, you'll learn how to manage numeric variables with missing values.

Handling missing values for numerical variables

In this section, you'll follow a systematic approach to handle missing values within numerical variables. You'll use descriptive statistics to fill in missing values, as well as learning how to use binary encoding techniques.

Identifying numerical variables with missing data

The first step in handling missing numeric values is to identify the affected variables. You can do this by looking at the non-categorical columns in the dataset and using the isnull() method:

1. First, let's build a list of all the categorical variables by looking for the variables that aren't numbers. You can check whether a variable isn't a number by looking for Python objects. Python objects have dtype == 'O' in them:

    ```
    vars_cat = [var for var in data.columns if data[
        var].dtype == "O"]
    ```

2. Let's get started by getting a count of all the numeric variables in the dataset. You can do this by counting the columns that aren't a member of the vars_cat list and aren't the target variable of "SalePrice". To do this, use a for loop to step through the X_train.colmns list and build a new list called vars_num that includes the columns not in vars_cat and not equal to "SalePrice". You can use len(vars_num) to get the number of numeric columns:

    ```
    vars_num = [var for var in X_train.columns if var not in
    vars_cat and var != "SalePrice"]
    len(vars_num)
    ```

This outputs the following:

    35

3. Now that you have a list of the numeric columns, you can check them for missing values using `.isnull().sum() > 0`. You can use this to build a list of the numeric variables with missing values called `vars_num_na`. To show the proportion of each column that's missing, you can use the `.mean()` method, like so:

```
vars_num_na = [var for var in vars_num if X_train[var].isnull().
sum() > 0]
X_train[vars_num_na ].isnull().mean()
```

This results in the following output:

```
LotFrontage    0.177321
MasVnrArea     0.004566
GarageYrBlt    0.056317
dtype: float64
```

Binary encoding for missing data and imputation with mean values

Before making changes so that you can fill in missing values, you can create an auxiliary binary variable for each numerical variable that has missing data. This approach enables you to preserve information about the proportion of missing data while also enabling imputation. By encoding how much data was missing for a variable as a binary indicator, you augment your dataset with additional information, which makes subsequent analyses more robust.

With the auxiliary binary variables in place, you can proceed to impute missing values within the original numerical variables. Using the mean value as a proxy for missing data, you can replace missing values with the calculated mean. This imputation strategy uses the central tendency of the data distribution, ensuring minimal disruption to the overall structure of your data's distribution. Let's take a look:

1. To create binary auxiliary columns that maintain information about where you originally had missing data, make a new column called `variable_na`, where *variable* is the name of the column with the missing values. This is set to 1 if the column has missing values and 0 if not. The best way to do this is with a `for` loop that steps through the list of numeric columns you made earlier, `vars_num_na`. While you're in that loop, you can also calculate the mean value for the columns with the `.mean()` method, putting the result in `mean_val`. As you did previously, you'll use the mean of the training dataset columns to replace values in both the training and test datasets. You can use the `.fillna` method as you did before. This looks like this:

```
for var in vars_num_na:
    # calculate the mean value for each variable from
    #the train set
    mean_val = X_train[var].mean()
    # creating binary indicator variable for
    #both  train and test set
    X_train[var + "_na"] = np.where(
        X_train[var].isnull(), 1, 0)
```

```
X_test[var + "_na"] = np.where(
    X_test[var].isnull(), 1, 0)
# replace missing values in the train and test set
#with train mean value
X_train[var].fillna(mean_val, inplace=True)
X_test[var].fillna(mean_val, inplace=True)
```

2. Now, you can perform a check, as you did for the categorical variables, to ensure you've filled in all the missing values. First, you can check `X_train` with `X_train[vars_num_na].isnull().sum()`. As before, if the sum is zero, all the missing values have been filled in:

    ```
    X_train[vars_num_na].isnull().sum()
    ```

    This produces the following output:

    ```
    LotFrontage    0
    MasVnrArea     0
    GarageYrBlt    0
    dtype: int64
    ```

3. Then, you can check `X_test` with `X_test[vars_num_na].isnull().sum()`:

    ```
    X_test[vars_num_na].isnull().sum()
    ```

    This gives us the following result:

    ```
    LotFrontage    0
    MasVnrArea     0
    GarageYrBlt    0
    dtype: int64
    ```

4. To wrap up this section, take a look at the binary indicator variables – that is, `LotFrontage_na`, `MasVnrArea_na`, and `GarageYrBlt_na` – using the `.head()` method:

    ```
    X_train[["LotFrontage_na", "MasVnrArea_na",
        "GarageYrBlt_na"]].head()
    ```

    This results in the following output:

	LotFrontage_na	MasVnrArea_na	GarageYrBlt_na
930	0	0	0
656	0	0	0
45	0	0	0
1348	1	0	0
55	0	0	0

Figure 7.2 – Result of X_train[["LotFrontage_na", "MasVnrArea_na", "GarageYrBlt_na"]].head()

With that, you've learned how to manage missing data for numeric variables. Next, you'll learn about feature engineering techniques you can implement for numeric variables. Transformations help make modeling easier – for example, if you can transform a variable into a Gaussian or normal distribution, this requires less math, and some models, such as linear regression, are built on the assumption that the data being fit is normal. That said, XGBoost doesn't make this assumption and will work even if you don't transform the data.

Common feature engineering techniques for numerical features

In the previous section, you learned how to handle missing data. In this section, you'll learn about feature engineering techniques that are commonly applied to numeric variables. These techniques do things such as transform data so that it's more conducive to modeling. Let's get started.

Performing a log transform

The log transformation is a powerful technique, particularly for transforming a variable into a normal distribution. In this section, you'll apply log transformation to positive numerical variables that exhibit non-normal distributions within the house prices dataset. You can refer to the data description provided on Kaggle for a comprehensive understanding of the variables that are subjected to log transformation. Let's get started:

1. To get started, plot the variables you'll transform to see how they don't follow a normal distribution initially. You can use `matplotlib` to plot these variables using a histogram to see their shape:

   ```
   import matplotlib.pyplot as plt
   ```

2. The variables of interest are `"LotFrontage"`, `"1stFlrSF"`, `"GrLivArea"`, and `"LotArea"`, so put them into a list called `var_for_log_trans`. You'll use this list later. You can plot histograms of these variables to see their shape. To make it easier to compare the variables, use `plt.subplots` to make a 1x4 array of graphs, then make the plots within a `for` loop. You can set the title for each graph so that it matches the plot by using `set_title(f"Histogram of {column}")` and `set_xlabel (column)` to set the X-axis label. To match the Y-axis range, use `set_ylim(0,1300)`. You can use `plt.tight_layout()` to control the spacing between the graphs. Once everything has been set up, use `plt.show()` to display the graphs:

   ```
   var_for_log_trans = ["LotFrontage", "1stFlrSF",
       "GrLivArea","LotArea"]
   plotdf = X_train [var_for_log_trans]
   fig, axes = plt.subplots(1,4, figsize=(16,4))
   ```

```
axes_flat = axes.flatten()

for i, (column, ax) in enumerate(zip(plotdf.columns,
    axes_flat)):
        plotdf[column].plot.hist(ax=ax, bins= 15,
            alpha = 0.5, color = "grey",
            edgecolor = "black")
    ax.set_title(f"Histogram of {column}",
        fontsize = 10)
    ax.set_xlabel(column, fontsize = 10)
    ax.set_ylim(0, 1300)

plt.tight_layout()
plt.show()
```

This results in the following graphs:

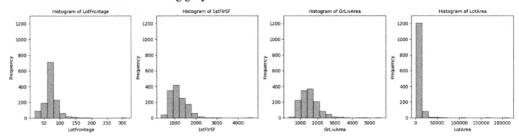

Figure 7.3 – Histograms of the skewed numeric columns

In the histograms shown in *Figure 7.3*, each variable is skewed or off-center, and some, such as LotArea, have a long tail, meaning there are a few rows with very large values for LotArea. This makes sense as generally, housing lots are typically measured in fractions of an acre, and there can be some that are multiple acres. The data in this dataset is measured in square feet, and an acre is 43,560 square feet. To make these variables more normal, you can apply a log transformation.

Log transformation rationale

The decision to perform log transformation stems from the observation of non-normal distributions within certain positive numerical variables. By applying log transformation, you can remove the **skewness** and **kurtosis** present in these distributions, matching them to the Gaussian distribution. This transformation not only improves the adherence to statistical assumptions but also enhances the interpretability and stability of subsequent analyses and predictive models. Let's perform this transformation on "LotFrontage", "1stFlrSF", "GrLivArea", and "LotArea":

1. Begin by looping over the variables, var, in var_for_log_trans. For each variable on the list in both X_train and X_test, calculate the log using np.log. You can put the result in a new column called "log_" + var:

```
for var in var_for_log_trans:
    X_train["log_"+ var] = np.log(X_train[var])
    X_test["log_" + var] = np.log(X_test[var])
```

2. Next, plot the transformed columns using log_trans = ["log_LotFrontage", "log_1stFlrSF", "log_GrLivArea", "log_LotArea"] to see how the distributions have changed, just as you did previously:

```
log_trans = ["log_LotFrontage", "log_1stFlrSF",
    "log_GrLivArea","log_LotArea"]
plotdf = X_train [log_trans]
fig, axes = plt.subplots(1,4, figsize=(16,4))
axes_flat = axes.flatten()

for i, (column, ax) in enumerate(zip(plotdf.columns,
    axes_flat)):
        plotdf[column].plot.hist(ax=ax, bins= 15,
            alpha = 0.5, color = "grey",
            edgecolor = "black")
    ax.set_title(f"Histogram of {column}",
        fontsize = 10)
    ax.set_xlabel(column, fontsize = 10)
    ax.set_ylim(0, 1300)

plt.tight_layout()
plt.show()
```

This results in the following graphs:

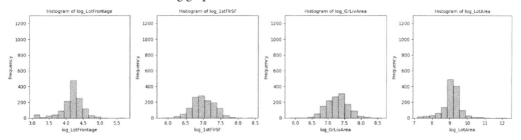

Figure 7.4 – Histograms of the skewed numeric columns after log transformation

3. After performing log transformation, the distributions shown in *Figure 7.4* follow the normal distribution. They are more centered on the *X*-axis and show the classic bell-shaped curve. Because of this benefit, let's also perform the log transformation on the target variable, `SalePrice`, which is in `y_train` and `y_test`:

```
y_train_log = np.log(y_train)
y_test_log = np.log(y_test)
```

Now that you've transformed skewed variables using a log transformation, let's discuss how to handle variables where there's one value that takes over and causes extreme skewing. In the house prices dataset, there are two of these: `ScreenPorch` and `EnclosedPorch`.

Handling skewed variables by transforming them into binary variables

Skewed variables can introduce bias and undermine the accuracy of predictive models. In this section, you'll learn how to transform variables where there's one value that takes over, causing extreme skewing, into binary variables. Skewed variables have asymmetrical distributions and can pose challenges in data analysis and modeling. You can identify these variables through data exploration.

In the case of the house prices dataset, two skewed variables will benefit from being converted into binary variables (for example false/true or no/yes): `ScreenPorch` and `EnclosedPorch`. The dataset contains multiple values for these variables. However, in most cases, they're zero, indicating that the house doesn't have a screened-in porch or doesn't have an enclosed porch. Because you know about houses – you have domain knowledge – you understand that knowing whether a house has or doesn't have a screened-in porch or an enclosed porch is a more important factor to the house price than specific details about the size of the screened-in or enclosed porch. By this reasoning, you can transform these columns so that they're binary: true or false. This will address the extreme skewness of these two columns. To see the skewness, let's make histogram graphs of `ScreenPorch` and `EnclosedPorch`:

1. You can use the same method you used to make the prior histograms. First, make a list of the columns you want to graph, called `skewed_vars = ['ScreenPorch', 'EnclosedPorch']`. This time, however, there are only two variables, so you can set up a matrix of plots to be one row with two plots, `fig, axes = plt.subplots(1,2, figsize=(16,4))`. Then, loop through the columns in the same way you did before and make the histograms with `plotdf[column].plot.hist`. Set the title for each graph inside the loop with `ax.set_title(f"Histogram of {column}", fontsize = 10)`. Set the labels for the *X*-axis with `ax.set_xlabel(column, fontsize = 10)` and the range of the *Y*-axis with `ax.set_ylim(0,1300)`:

```
skewed_vars = ['ScreenPorch', 'EnclosedPorch']
plotdf = X_train [skewed_vars]
fig, axes = plt.subplots(1,2, figsize=(16,4))
```

```
axes_flat = axes.flatten()

for i, (column, ax) in enumerate(zip(plotdf.columns,
    axes_flat)):
        plotdf[column].plot.hist(ax=ax, bins= 15,
        alpha = 0.5, color = "grey",
        edgecolor = "black")
    ax.set_title(f"Histogram of {column}",
        fontsize = 10)
    ax.set_xlabel(column, fontsize = 10)
    ax.set_ylim(0, 1300)

plt.tight_layout()
plt.show()
```

This results in the following output:

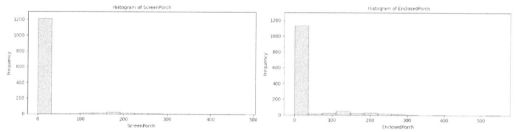

Figure 7.5 – Extremely skewed variables – ScreenPorch and EnclosedPorch

The histograms shown in *Figure 7.5* illustrate what's meant by extreme skewness. There are many more instances of rows with zero values in these two columns than there are rows with non-zero values. Next, let's transform these into binary variables.

2. To transform the ScreenPorch and EnclosedPorch variables into binary, you need to make a rule that says that if the value is zero, keep it as zero, but if the value is anything else, set it to one. You should do this in a new column so that you can keep the actual values for future analysis. Create a for loop over var in the skewed_vars list, and then do the mapping for both X_train and X_test:

```
for var in skewed_vars:
    # map the variable values into 0 and 1
    X_train["binary_"+ var] = np.where(
        X_train[var]==0, 0, 1)
    X_test["binary_"+ var] = np.where(
        X_test[var]==0, 0, 1)
```

Now, when you plot the histograms for `ScreenPorch` and `EnclosedPorch`, you'll see a bar at 0 and a bar at 1 for each variable. We'll do that next.

3. You can plot the transformed columns, `binary_trans = ["binary_ScreenPorch", "binary_EnclosedPorch"]`, to see how the distributions have changed, just as you did previously:

```
binary_trans = ["binary_ScreenPorch",
    "binary_EnclosedPorch"]
plotdf = X_train [binary_trans]
fig, axes = plt.subplots(1,2, figsize=(16,4))
axes_flat = axes.flatten()

for i, (column, ax) in enumerate(zip(plotdf.columns,
    axes_flat)):
        plotdf[column].plot.hist(ax=ax, bins= 15,
            alpha = 0.5, color = "grey",
            edgecolor = "black")
    ax.set_title(f"Histogram of {column}",
        fontsize = 10)
    ax.set_xlabel(column, fontsize = 10)
    ax.set_ylim(0, 1300)

plt.tight_layout()
plt.show()
```

This results in the following histograms:

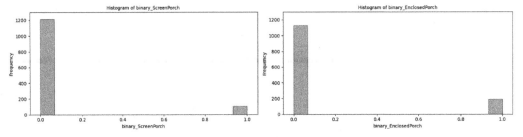

Figure 7.6 – Extremely skewed variables after transformation

In this section, you learned how transforming skewed variables into binary variables offers a practical strategy for handling extreme skewness. You were able to achieve a more balanced representation of the underlying data distribution, thereby facilitating accurate and reliable model predictions. Next, you'll learn how to scale features so that they align with a common scale.

Scaling features

Scaling features is a preprocessing step that's used in many machine learning algorithms, particularly for linear models, where it can significantly impact model performance. In this section, you'll learn about the importance of scaling features and learn about min-max scaling, a commonly used scaling technique.

Understanding feature scaling

Feature scaling involves transforming numerical features into a common scale, typically between 0 and 1, to ensure uniformity and comparability across features. This normalization process prevents features with larger scales from dominating those with smaller scales during model training, which improves model convergence and performance. While scaling doesn't impact decision-tree-based model performance, linear models and distance-based algorithms greatly benefit from scaled features. In this section, you'll use normalization, also known as min-max scaling, to rescale features to a specified range. This scaling technique preserves the relative relationships between feature values while compressing the range of values to a uniform scale. Min-max scaling is particularly suitable for features with bounded ranges and is widely used in various machine learning algorithms.

Implementing min-max scaling

Implementing min-max scaling involves transforming feature values according to the following formula:

$$X_{scaled} = \frac{(X - X_{min})}{(X - X_{max})}$$

Here, X_{min} and X_{max} denote the minimum and maximum values of the feature, respectively. This transformation is applied independently to each feature, ensuring consistency and preserving the integrity of the data. Let's get started:

1. First, create a new DataFrame called `X_train_num` that only contains the numeric columns, `vars_num`. This prevents errors due to us attempting to perform statical calculations on a string or other non-number value:

   ```
   X_train_num = X_train[vars_num]
   ```

2. You can look at the values using `X_train_num.head()`:

   ```
   X_train_num.head()
   ```

   This produces the following result:

	LotFrontage	LotArea	OverallQual	OverallCond	YearBuilt	YearRemodAdd	MasVnrArea	BsmtFinSF1	BsmtFinSF2	BsmtUnfSF	TotalBsmtSF
930	73.000000	8925	8	5	2007	2007	0.0	16	0	1450	1466
656	72.000000	10007	5	7	1959	2006	54.0	806	0	247	1053
45	61.000000	7658	9	5	2005	2005	412.0	456	0	1296	1752
1348	69.879741	16196	7	5	1998	1998	0.0	1443	0	39	1482
55	100.000000	10175	6	5	1964	1964	272.0	490	0	935	1425

Figure 7.7 – Result of X_train_num.head()

3. Let's do the same thing for `X_test`, creating a DataFrame `X_test_num` with just the numeric columns:

```
X_test_num = X_test[vars_num]
X_test_num.head()
```

This produces the following result:

	LotFrontage	LotArea	OverallQual	OverallCond	YearBuilt	YearRemodAdd	MasVnrArea	BsmtFinSF1	BsmtFinSF2	BsmtUnfSF	TotalBsmtSF
529	69.879741	32668	6	3	1957	1975	103.797401	1219	0	816	2035
491	79.000000	9490	6	7	1941	1950	0.000000	403	165	238	806
459	69.879741	7015	5	4	1950	1950	161.000000	185	0	524	709
279	83.000000	10005	7	5	1977	1977	299.000000	392	0	768	1160
655	21.000000	1680	6	5	1971	1971	381.000000	0	0	525	525

Figure 7.8 – Result of X_test_num.head()

4. Now, you have only the numeric columns in both `X_train_num` and `X_test_num` and can use the `MinMaxScaler` module from `sklearn.preprocessing` to scale the values. To begin, import the package and create a scaler by calling `MinMaxScaler()`:

```
from sklearn.preprocessing import MinMaxScaler
set_scale = MinMaxScaler()
```

5. Next, fit the scaler using the training dataset. You'll use this same scaler on both `X_train_num` and `X_test_num`. You can name the fitted scaler, `scaler`:

```
scaler = set_scale.fit(X_train_num.values)
```

6. With this fitted scaler, transform both the training and testing datasets. By default, `MinMaxScaler` produces an `np.array` value. To keep the output as a DataFrame, you need to call `pd.DataFrame` with the values, columns, and index for `X_train_num` and `X_test_num`, like so:

```
X_train_scale = pd.DataFrame(scaler.transform(
    X_train_num.values), columns=X_train_num.columns,
    index=X_train_num.index)
X_test_scale = pd.DataFrame(scaler.transform(
    X_test_num.values), columns=X_test_num.columns,
    index=X_test_num.index)
```

7. Now, you can look at the first few rows after the transformation to see what happened:

```
X_train_scale.head()
```

This prints the following:

	LotFrontage	LotArea	OverallQual	OverallCond	YearBuilt	YearRemodAdd	MasVnrArea	BsmtFinSF1	BsmtFinSF2	BsmtUnfSF	TotalBsmtSF
930	0.178082	0.035640	0.777778	0.50	0.978261	0.950000	0.00000	0.002835	0.0	0.673479	0.239935
656	0.174658	0.040697	0.444444	0.75	0.630435	0.933333	0.03375	0.142807	0.0	0.114724	0.172340
45	0.136986	0.029718	0.888889	0.50	0.963768	0.916667	0.25750	0.080794	0.0	0.601951	0.286743
1348	0.167396	0.069625	0.666667	0.50	0.913043	0.800000	0.00000	0.255670	0.0	0.018114	0.242553
55	0.270548	0.041483	0.555556	0.50	0.666667	0.233333	0.17000	0.086818	0.0	0.434278	0.233224

Figure 7.9 – Result of X_train_scale.head()

Comparing *Figure 7.7*, which shows these columns before the transformation, to *Figure 7.9*, which shows the result, allows you to see what's happened. Before the scaling, there was a wide variety of values in the various columns. Some were single digits, while some were in the thousands. After performing scaling, all the values have been remapped to the 0 to 1 range.

In this section, you learned about common featuring engineering techniques for numeric variables. You handled missing data using binary encoding and imputation. You also transformed skewed variables by either performing a log transform or by converting them into binary variables in the case of extreme skewness. Finally, you used min-max scaling to transform data from a wide range of values to a common range, something that allows you to make easier comparisons across variables. Next, you'll learn about common feature engineering techniques for time and date data.

Common feature engineering techniques for temporal features

In this section, you'll learn about feature engineering techniques that are commonly applied to time and date variables. You'll learn how to handle missing data as well as create additional useful time and date columns (feature engineering). Let's begin by looking at how to manage missing temporal data.

Handling temporal variables

Temporal variables represent time-related attributes within a dataset and present unique challenges. In this section, you'll explore strategies for handling temporal variables within the context of the house prices dataset. Specifically, you'll focus on three key temporal variables: GarageYrBlt (year garage was built), YearBuilt (original construction date), and YearRemodAdd (date of a remodel or addition). Looking at these variables, you can derive meaningful insights into the dynamics of housing construction and renovation. Understanding the time-related dynamics in these variables can show patterns and trends that shape the housing market landscape.

Deriving time differences

To begin extracting actionable insights from temporal variables, you need to engineer some additional columns that look at the difference in time between each time-based variable and the year in which the house was sold. This will help you see the evolution of housing construction and renovation practices over time. This approach allows you to identify time-based correlations and provides valuable context for predictive modeling. Let's get started by creating a function that calculates the time difference between a variable that you provide and the year the house was sold:

1. Define a function called `years_diff` that performs the calculation by taking the difference between the date you pass to the function, `var`, and the `YrSold` column in `dataframe` `df["YrSold"]`. You can create a new column to put the result in, call it `YrSold_-_` *variable*, and return the `df` DataFrame at the end of the function. This allows you to keep the original values in case your calculation doesn't do what you want it to:

    ```
    def years_diff(df, var):
        df["YrSold_-_" + var] = df["YrSold"] - df[var]
        return df
    ```

2. Now, let's use this function to make new columns with the date difference between `YrSold` and `YearBuild`, `YearRemodAdd`, and `GarageYrBlt`. You can make this easy by using a `for` loop and stepping through a list of the columns you want to do the calculation on. Since the function replaces the values in the columns you pass with the result of the date difference calculation, you'll want to put the results into new DataFrames – that is, `X_train_diff` and `X_test_diff`:

    ```
    for var in ["YearBuilt", "YearRemodAdd",
        "GarageYrBlt"]:
            X_train_diff = years_diff(X_train, var)
            X_test_diff = years_diff(X_test, var)
    ```

3. At this point, you'll want to check that the calculation produces the result you expect. You can look at a single row for the particular columns by printing out the row and variables you wish using `.loc`. This example uses `X_train.loc[0,"YrSold"]` to get the value in the `YrSold` column for the row of the table at index 0, and `X_train.loc[0,"YearBuilt"]`, `X_train.loc[0,"YearRemodAdd"]`, `X_train.loc[0,"GarageYrBlt"]` to get the original values in the `YearBuild`, `YearRemodAdd`, and `GarageYrBlt` columns for the same row. Put that in a `print` statement with the `"Original table: "` label and repeat this for `"Date difference"` using `X_train_diff.loc` to show the before and after of the date difference calculation, like so:

    ```
    print("Original table: ", X_train.loc[0,"YrSold"],
        X_train.loc[0,"YearBuilt"], X_train.loc[0,"YearRemodAdd"],
        X_train.loc[0,"GarageYrBlt"])
    ```

```
print("Date difference: ", X_train_diff.loc[0,"YrSold"],
X_train_diff.loc[0,"YrSold_-_YearBuilt"], X_train_diff.
loc[0,"YrSold_-_YearRemodAdd"], X_train_diff.loc[
    0,"YrSold_-_GarageYrBlt"])
```

This produces the following output:

```
Original table:   2008 2003 2003 2003.0
Date difference:  2008 5 5 5.0
```

4. Since the `YrSold` variable is now encoded in the date differences to `YearBuild`, `YearRemodAdd`, and `GarageYrBlt`, you can drop `YrSold` from both the `X_train_diff` and `X_test_diff` datasets:

```
X_train_diff.drop(["YrSold"], axis=1, inplace=True)
X_test_diff.drop(["YrSold"], axis=1, inplace=True
```

Now, you have three new columns: `YrSold_-_YearBuilt`, `YrSold_-_YearRemodAdd`, and `YrSold_-_GarageYrBlt`. These columns show the difference in years between the date the house was sold and when it was built, which is the age of the house when it was sold, the age when it was remodeled, and the age when a garage was built.

5. You can see the new columns that have been added to the far right of the DataFrame by using `X_train_diff.head()` and scrolling to the right with the scroll bar, as shown in *Figure 7.3*:

```
X_train_diff.head()
```

scFeature	MiscVal	MoSold	SaleType	SaleCondition	LotFrontage_na	MasVnrArea_na	GarageYrBlt_na	YrSold_-_YearBuilt	YrSold_-_YearRemodAdd	YrSold_-_GarageYrBlt
Missing	0	7	WD	Normal	0	0	0	2	2	2.0
Missing	0	8	WD	Normal	0	0	0	49	2	49.0
Missing	0	2	WD	Normal	0	0	0	5	5	5.0
Missing	0	8	WD	Normal	1	0	0	9	9	9.0
Missing	0	7	WD	Normal	0	0	0	44	44	44.0

Figure 7.10 – Result of X_train_diff.head()

Now, you have three new columns that calculate the age at which houses have been remodeled or had an addition or a garage, as well as the age at which a house was sold. These columns give you additional information by comparing the date of these events to the date the house was built.

This section covered common techniques for engineering features from date and time data. Next, you'll learn about common feature engineering techniques for categorical variables.

Common feature engineering techniques for categorical features

Categorical variables often contain useful information that can't be easily quantified. In this section, you'll learn about methods for transforming categorical variables to enhance their utility and interpretability in predictive modeling. Specifically, you'll learn about mappings based on inherent order or related quality within categorical variables.

Understanding categorical variables

There are two common types of categorical variables: ordinal and nominal. Ordinal variables have a specific order or ranking. This would include things such as T-shirt sizes, where small comes first, then medium, followed by large, or ranking of things such as our tomato example of poor, fair, good, and excellent.

Nominal variables, on the other hand, are unordered categories. This would include things such as names of defects, or types of parts. Transforming categorical variables involves encoding these attributes into a numerical format so that they can be used in statistical analysis and modeling.

Let's begin by handling ordinal values with order mappings. Using this method, you can capture the ordinal nature or related quality within categorical variables, thereby enriching the information content and predictive power of the dataset.

Order mappings – capturing order or related quality

Order mappings assign numerical values to categorical variables based on their inherent order or related quality. This transformation preserves the ordinal relationship or qualitative hierarchy within categorical variables, facilitating meaningful comparisons and interpretations in predictive modeling contexts. Whether you're mapping ordinal rankings, quality ratings, or hierarchical levels, order mappings imbue categorical variables with a structured numerical representation that aligns with their underlying semantics. In the house prices dataset, there are several variables with ordinal values. Please refer to the data description for details: `https://www.kaggle.com/competitions/house-prices-advanced-regression-techniques/data?select=data_description.txt`. You'll map all the ordinal variables in this section.

Mapping variables with quality assessments of fair to excellent

First, you can map qualitative assessments such as "poor," "fair," "good," and "excellent" to numbers such as 1 = "poor" to 5 = "excellent." Let's start with the nine variables that assess quality in the dataset:

1. Start by making a dictionary that associates each string with a value. Don't forget to include a value for missing or NA. In this case, you can map those choices to 0. You can call the dictionary map_quality:

    ```
    map_quality = {"Po": 1, "Fa": 2, "TA": 3, "Gd": 4,
        "Ex": 5, "Missing": 0, "NA": 0}
    ```

2. Put the variables you wish to encode into a list. In this case, you'll work on ExterQual, ExterCond, BsmtQual, BsmtCond, HeatingQC, KitchenQual, FireplaceQu, GarageQual, and GarageCond:

    ```
    vars_qual = ["ExterQual", "ExterCond", "BsmtQual",
                "BsmtCond", "HeatingQC", "KitchenQual",
                "FireplaceQu", "GarageQual","GarageCond",
                ]
    ```

3. Use the dictionary to map the ordinal assessment string to a quantity and put the result in a new column called "map_" *variable* for both the X_train and X_test DataFrames:

    ```
    for var in vars_qual:
        X_train["map_"+ var] = X_train[var].map(
            map_quality)
        X_test["map_"+ var] = X_test[var].map(map_quality)
    ```

    This results in new columns with numeric representations of the string quantitative assessment of various qualities of a house.

4. You can get a sense of the changes by looking at the first few rows for a subset of the modified columns with X_train[["ExterQual", "map_ExterQual","BsmtQual","map_BsmtQual","FireplaceQu","map_FireplaceQu"]].head():

    ```
    X_train[["ExterQual", "map_ExterQual","BsmtQual",
        "map_BsmtQual","FireplaceQu",
        "map_FireplaceQu"]].head()
    ```

    This gives us the following output:

	ExterQual	map_ExterQual	BsmtQual	map_BsmtQual	FireplaceQu	map_FireplaceQu
930	Gd	4	Gd	4	Missing	0
656	Gd	4	TA	3	Missing	0
45	Ex	5	Ex	5	Gd	4
1348	Gd	4	Gd	4	Fa	2
55	TA	3	TA	3	Gd	4

Figure 7.11 – Result of mapping values to ordinal variables

Mapping other variables with ordered values

You can apply this same concept to variables that have an order but don't follow the same classifications. For each classification scheme, you'll need to create a dictionary to associate numbers with the values. Follow these steps:

1. Apply the same mapping concept to `BsmtExposure` and convert it from a string into numeric values:

```
map_exposure = {'No': 1, 'Mn': 2, 'Av': 3, 'Gd': 4}
var_expose = 'BsmtExposure'
X_train["map_"+ var_expose] = X_train[
    var_expose].map(map_exposure)
X_test["map_"+ var_expose] = X_test[
    var_expose].map(map_exposure)
```

2. Next, map the two columns related to the finished basements. You need to handle each variable with different assessment values separately, but you can group columns that use the same values together and convert them at the same time using the same dictionary:

```
map_finish = {'Missing': 0, 'NA': 0, 'Unf': 1,
    'LwQ': 2, 'Rec': 3, 'BLQ': 4, 'ALQ': 5, 'GLQ': 6}
vars_finish = ['BsmtFinType1', 'BsmtFinType2']
for var in vars_finish:
    X_train["map_"+ var]= X_train[var].map(map_finish)
    X_test["map_"+ var] = X_test[var].map(map_finish)
```

3. Next, you can convert the `GarageFinish` column:

```
map_garage = {'Missing': 0, 'NA': 0, 'Unf': 1,
    'RFn': 2, 'Fin': 3}
var_garage = 'GarageFinish'
X_train["map_"+ var_garage] = X_train[
    var_garage].map(map_garage )
X_test["map_"+ var_garage] = X_test[
    var_garage].map(map_garage )
```

4. The final ordinal column that needs to be mapped is `Fence`:

```
map_fence = {'Missing': 0, 'NA': 0, 'MnWw': 1,
    'GdWo': 2, 'MnPrv': 3, 'GdPrv': 4}
var_fence = 'Fence'
X_train["map_"+ var_fence] = X_train[
    var_fence].map(map_fence)
X_test["map_"+ var_fence] = X_test[
    var_fence].map(map_fence)
```

In this section, you used order mapping to apply a dictionary that associates strings with an ordered numeric value. You applied this mapping to columns where there was a qualitative assessment and a ranking in the associated values and used dictionaries to associate numbers with the various strings that classified the rankings. This mapping converts the strings into numbers, which can be used in model training. In the next section, you'll handle nominal variables. These are categorical variables that don't have a specific order associated with them.

Identifying rare labels for the rest of the categorical/nominal variables

As mentioned previously, nominal variables are categorical variables that don't have a specific order associated with them. This includes things such as the names of defects, or types of parts. Transforming categorical variables involves encoding these attributes into a numerical format so that they can be used in statistical analysis and modeling. You can start by identifying labels that aren't common within the dataset – that is, rare labels.

Identifying and handling rare labels within categorical variables is a critical preprocessing step in predictive modeling. By consolidating rare labels and standardizing their representation, we can mitigate the risk of overfitting and improve the generalizability of predictive models.

Categorical variables often harbor a multitude of unique values, ranging from common categories to rare labels that occur infrequently within the dataset. In this section, you'll use a methodology for identifying and handling rare labels within categorical variables. By discerning and consolidating rare labels, you can streamline the categorical variable space and enhance the robustness of predictive modeling endeavors.

Rare labels represent categories within categorical variables that occur infrequently, typically accounting for a small proportion of observations. These rare labels, while valid representations of the underlying data, pose challenges in predictive modeling due to their limited occurrence. Identifying and handling rare labels is essential for mitigating the risk of overfitting and improving the generalizability of predictive models.

To identify rare labels within categorical variables, you can use a threshold criterion based on frequency of occurrence. Specifically, you can designate labels that are shared by less than 1% of houses as rare. All the values of categorical variables that qualify as rare, as per the designated threshold, are replaced with the `"Rare"` string. This consolidation of rare labels simplifies the categorical variable space and minimizes the risk of overfitting associated with sparse categories. By standardizing the representation of rare labels, you enhance the interpretability and predictive power of the dataset. Let's get started:

1. Begin by creating a list of all the categorical variables you've already mapped:

    ```
    vars_qual_all = vars_qual + [var_expose] + vars_finish + [var_
    garage] + [var_fence]
    ```

2. Now, you can create a list that contains the rest of the categorical variables by selecting the `cat_others` variables from the list of categorical variables, `vars_cat`, that aren't in the list of already mapped quality measurements, `vars_qual_all`:

    ```
    cat_others = [ var for var in vars_cat if var not in vars_qual_
    all ]
    ```

3. Let's look at how many variables fall into this category:

    ```
    len(cat_others)
    ```

    This produces the following output:

    ```
    30
    ```

4. Now, you can build a function that finds the rare labels in the columns in `cat_others`. This function takes the dataset to look at, `df`, which columns to assess, `var`, and the threshold for defining a rare value, `perc`. Then, it uses pandas' grouping ability to create a count for each possible value, `df.groupby(var)[var].count()`, and assess the number of items with that value against the threshold you set for a rare value:

    ```
    # this function will find rare labels
    def find_rare_labels(df, var, perc):
        df_copy = df.copy()
        # find the counts per categorical value
        tmp_cat = df.groupby(var)[var].count() / len(df)
        # return categories that very rare
        return tmp_cat[tmp_cat < perc].index
    ```

5. Now, you can apply the function to find the rare labels within the dataset using a threshold of 0.01 or 1%. To do this, call the function and send it `X_train`, `var`, and `0.01`, putting the results into `rare_arr`. Then, you can print `rare_arr`:

    ```
    for var in cat_others:
        # find the rare labels list
        rare_arr = find_rare_labels(X_train, var, 0.01)
    ```

```
print(var, rare_arr)
print()
# replace rare categories by the string "Rare"
X_train["rare_" + var] = np.where(
    X_train[var].isin(rare_arr), 'Rare', X_train[
        var])

#remember to use the same for the test set
X_test["rare_" + var] = np.where(X_test[var].isin(
    rare_arr), 'Rare', X_test[var])
```

This prints the following:

```
MSZoning Index(['C (all)'], dtype='object', name='MSZoning')

Street Index(['Grvl'], dtype='object', name='Street')

Alley Index([], dtype='object', name='Alley')

LotShape Index(['IR3'], dtype='object', name='LotShape')

. . .
```

> **Note**
> The preceding output has been truncated for brevity. For the full output, please refer to the respective Jupyter notebook in this book's GitHub repository.

With that, you've marked all the values with a "rare" label if they're infrequent in the dataset. Having done that, you can now encode the remaining classifications and names as numbers so that they can be used in further analysis. You'll do that for the house prices dataset in the next section.

Encoding categorical variables using a simple technique

To use categorical variables effectively in a machine learning model, it's helpful to be able to convert from a string into a numeric value. By mapping each category to a numerical representation, we can establish a structured framework for analyzing and modeling categorical variables. In this section, you'll explore a simple encoding technique while using the relationship between categorical labels and the target variable. More advanced encoding techniques that build on this method will be provided in *Chapter 8*.

In this simple implementation, you'll leverage the relationship between categorical labels and the target variable. Specifically, you'll map each category value to an integer based on the mean sale price, ranging from low to high. This ordinal encoding scheme captures the relationship between categories and their corresponding target variable values, facilitating intuitive interpretation and analysis:

1. First, create a function that will assign an integer to a category based on it being associated with the mean `SalePrice` ranked from low to high. To do this, you will need `X_train`, `X_test`, the variable to work on from the dataset, `var`, and the target variable, `target`, which is in `y_train`. In this function, sort the columns in order based on the target variable and put the result into `category_order` with `category_order = df_tmp.groupby([var])[target].mean().sort_values().index`. Then, use `enumerate` to associate a value from 1 to *n* for each of the items in `category_order`:

```python
def impute_category(X_train, X_test, y_train, var,
    target):
        df_tmp = pd.concat([X_train, y_train], axis=1)
        #create the list of categorical values for a
#variable sorted by the mean Sale Price less to high
    category_order = df_tmp.groupby(
        [var])[target].mean().sort_values().index
        # create mapping for each category value to an
#integer in increasing order
    ordered_num_labels = {k: i for i, k in enumerate(
        category_order)}
        #e.g.: {k: i for i, k in
        # enumerate(['a','b','c','d'])}
    # {'a': 0, 'b': 1, 'c': 2, 'd': 3}

    print(var, ordered_num_labels)
    print()

    # use the mapping to replace categories
    X_train["encode_"+ var] = X_train[
        var].map(ordered_num_labels)
    X_test["encode_" + var] = X_test[
        var].map(ordered_num_labels)
```

2. Apply the function and pass it `X_train`, `X_test`, `y_train`, `var`, and the target, `"SalePrice"`.

    This will produce the following output:

    ```
    MSZoning {'C (all)': 0, 'RM': 1, 'RH': 2, 'RL': 3, 'FV': 4}
    Street {'Grvl': 0, 'Pave': 1}
    Alley {'Grvl': 0, 'Pave': 1, 'Missing': 2}
    LotShape {'Reg': 0, 'IR1': 1, 'IR3': 2, 'IR2': 3}
    LandContour {'Bnk': 0, 'Lvl': 1, 'Low': 2, 'HLS': 3}
    Utilities {'NoSeWa': 0, 'AllPub': 1}
    LotConfig {'FR2': 0, 'Inside': 1, 'Corner': 2, 'FR3': 3,
    'CulDSac': 4}
    LandSlope {'Gtl': 0, 'Mod': 1, 'Sev': 2}
    Neighborhood {'IDOTRR': 0, 'MeadowV': 1, 'BrDale': 2, 'Edwards':
    3, 'BrkSide': 4, 'OldTown': 5, 'Sawyer': 6, 'Blueste': 7,
    'NPkVill': 8, 'SWISU': 9, 'NAmes': 10, 'Mitchel': 11, 'SawyerW':
    12, 'NWAmes': 13, 'Gilbert': 14, 'Blmngtn': 15, 'CollgCr': 16,
    'ClearCr': 17, 'Crawfor': 18, 'Somerst': 19, 'Veenker': 20,
    'Timber': 21, 'StoneBr': 22, 'NridgHt': 23, 'NoRidge': 24}
    ...
    ```

> **Note**
>
> The preceding output has been truncated for brevity. For the full output, please refer to the respective Jupyter notebook in this book's GitHub repository.

3. You can check for null values in the training set using the following code:

    ```
    [var for var in X_train.columns if X_train[
        var].isnull().sum() > 0]
    ```

    This should produce the following output:

    ```
    []
    ```

    This is a sanity check that ensures the encoding is working as expected. If it doesn't come back empty, this means we should check the logic or the code for the encoding technique.

4. For the testing dataset, you can use the following code:

    ```
    [var for var in X_test.columns if X_test[
        var].isnull().sum() > 0]
    ```

    This will display the following output:

    ```
    ['encode_Condition2']
    ```

5. You can look at the encoded variables by creating a temporary DataFrame that uses a filter and using `regex` to find the columns in `X_train` that start with `"encode_"` and concatenate `y_train` so that it includes `"SalePrice"`. You can use this DataFrame in the next step to make graphs. But first, just use `df_tmp.head()` to look at the first rows:

```
df_tmp = pd.concat([X_train.filter(regex='^encode_'),
    np.log(y_train)], axis=1)
df_tmp.head()
```

This will result in the following output for the first few columns:

	encode_MSZoning	encode_Street	encode_Alley	encode_LotShape	encode_LandContour	encode_Utilities	encode_LotConfig	encode_LandSlope
930	3	1	2	1	3	1	1	0
656	3	1	2	1	1	1	1	0
45	3	1	2	0	1	1	1	0
1348	3	1	2	2	2	1	1	0
55	3	1	2	1	1	1	1	0

Figure 7.12 – The first few columns and first rows of the temporary DataFrame

With that, each of these categorical variables has been ranked as numeric values, as shown in *Figure 7.12*.

6. Finish up this section by making plots of the encoded columns and the target variable, `SalePrice`. Since there are 30 encoded variables, you can use a loop to plot a few at a time and not have to scroll so much or deal with them being small and hard to view. The following code graphs five columns at a time in a loop. Feel free to play around and try other values:

```
import seaborn as sns

for i in range (0, len(cat_others),5):
    g = sns.pairplot(df_tmp, y_vars=["SalePrice"],
    x_vars=["encode_"+ cat_others[i],
        "encode_"+ cat_others[i+1],
        "encode_"+ cat_others[i+2],
        "encode_"+ cat_others[i+3],
        "encode_"+ cat_others[i+4]])
```

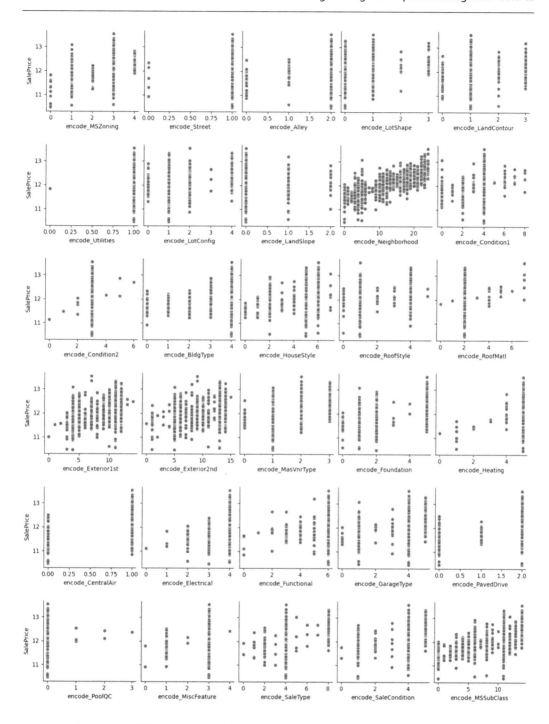

Figure 7.13 – Graphs of SalePrice versus the encoded categorical variables

Encoding categorical variables is a critical preprocessing step in predictive modeling that allows you to incorporate qualitative attributes into statistical models. By adopting a simple yet effective encoding technique based on the relationship between categorical labels and the target variable, you were able to establish a structured numerical representation that facilitates analysis and modeling. *Chapter 8*, will further develop categorical encoding techniques, which will allow you to develop further preprocessing tools for predictive modeling.

Summary

This chapter served as an in-depth exploration into fundamental feature engineering methodologies, complemented by practical demonstrations within Jupyter notebooks using the house prices dataset sourced from Kaggle. Through a systematic approach, you practiced using various feature engineering techniques that are useful for performing predictive modeling. You learned about handling missing values, transforming numeric data, and how to handle skewed data. You also performed feature engineering on temporal data, creating new columns that enable insights into the evolution of housing attributes over time.

A pivotal lesson that was underscored throughout this chapter is the importance of splitting data at the outset of your analysis. By segregating your data into distinct training and testing sets before feature engineering, you safeguard against data leakage and uphold the integrity of your model evaluation process. This foundational step lays the groundwork for robust model development and validation.

In conclusion, this chapter has equipped you with a comprehensive understanding of essential feature engineering techniques, arming you with the tools and insights necessary to navigate complex datasets effectively. In *Chapter 8*, you'll explore additional categorical feature engineering and encoding techniques.

Encoding Techniques for Categorical Features

This chapter is a hands-on guide that provides you with practical experience of using various categorical encoding techniques. You will be implementing several categorical encoding techniques using Python, specifically with the pandas and scikit-learn libraries.

Categorical encoding is the process of converting categorical data (i.e., non-numeric data representing categories or labels) into numerical formats that **machine learning** (**ML**) algorithms can use. Most ML models and statistical techniques require numerical input, making it necessary to transform categorical variables into a form that the model can understand and process.

In *Chapter 7*, you learned how to handle missing values for categorical variables. Additionally, that chapter covered the simple encoding techniques of order mapping and mapping nominal data, including how to handle rare categories. We introduced the two types of categorical data that these encoding techniques can be applied to: nominal and ordinal data. The techniques you will learn in this chapter can be used to effectively transform both types of categorical data into numerical formats, enabling their use in ML models.

In this chapter we're going to cover the following main topics:

- Why is categorical encoding necessary?
- One-hot encoding
- One-hot encoding with frequent categories
- Integer encoding
- Count or frequency encoding
- Ordered integer encoding
- Mean encoding
- Probability ratio encoding

- Weight of evidence (WoE) encoding
- Engineering rare categories

Technical requirements

To effectively follow along with the code examples and exercises in this chapter on categorical encoding techniques, you will need to have the following Python libraries. The code provided in this chapter consists of Python functions designed to implement various encoding techniques for categorical data. All code examples and datasets used in this chapter are available on our GitHub repository at `https://github.com/PacktPublishing/XGBoost-for-Regression-Pre-dictive-Modeling-and-Time-Series-Analysis/blob/main/ch8/categori-cal_encoding.ipynb`.

You will need the following Python packages:

- pandas 1.4.2
- NumPy 1.21.5
- Scikit-learn 1.2.2
- Matplotlib

As in *Chapter 7*, you will be using the *Housing Price* dataset. Access is available through the Kaggle platform at the following link: `https://www.kaggle.com/competitions/house-prices-advanced-regression-techniques/data`. To gain an understanding of the variables within the dataset, you can refer to the accompanying data description, accessible via the following link: `https://www.kaggle.com/competitions/house-prices-advanced-regression-techniques/data?select=data_description.txt`. You will be using this dataset in subsequent chapters as well.

> **Dataset citation**
>
> Dean De Cock (2011) Ames, Iowa: *Alternative to the Boston Housing Data as an End of Semester Regression Project*, Journal of Statistics Education, 19:3, DOI: 10.1080/10691898.2011.11889627.

Why is categorical encoding necessary?

The primary reason to perform categorical encoding is that ML algorithms require numerical input. Most ML algorithms are mathematical models that perform operations on numerical data. They calculate distances, optimize weights, and apply mathematical transformations that require numerical input. As a result, categorical data (text values) must be converted into numbers before being fed into these models. For example, in a logistic regression model, the algorithm needs to calculate the likelihood of an outcome based on input features. If these features are non-numeric, the model cannot perform the necessary calculations.

Another reason is for model interpretation and decision-making. For some models, such as decision trees, the ability to split data based on numeric thresholds is essential. Categorical data must be encoded numerically to allow the model to create meaningful splits and make decisions at each node of the tree. Encoding helps preserve the interpretability of the model by ensuring that the relationships between features and the target variable are accurately represented.

Categorical variables with a large number of categories (high cardinality) can be problematic because they can lead to overfitting or cause the model to be less efficient. Encoding techniques can help manage this by reducing the dimensionality or by grouping categories in a meaningful way. For example, a feature representing cities might have hundreds or thousands of unique categories. Applying encoding techniques such as one-hot encoding, where each category is turned into a column of data (see the *One-hot encoding* section for more), to such a feature would create an impractically large number of binary columns. Techniques such as target encoding or frequency encoding can help in reducing this complexity.

Some encoding techniques, such as mean encoding or **Weight of Evidence (WoE)** encoding, capture the relationship between a categorical feature and the target variable. This can improve the predictive power of the model by encoding categories in a way that reflects their influence on the target. For instance, mean encoding replaces each category with the mean of the target variable for that category, allowing the model to better capture patterns and trends in the data.

Proper encoding can help avoid bias in model predictions by ensuring that all categories are treated appropriately. For example, integer encoding may inadvertently introduce bias by implying a relationship between categories that does not exist. More sophisticated techniques such as one-hot encoding or target encoding can help mitigate such biases.

Lastly, the way categorical data is encoded can significantly impact the performance of an ML model. Incorrect or suboptimal encoding can lead to poor model performance, while appropriate encoding can enhance the model's ability to generalize to new data. For example, in a linear regression model, one-hot encoding can be used to avoid multicollinearity, which occurs when independent variables are highly correlated. This ensures that the model does not overfit or make inaccurate predictions.

To summarize, in this section, you learned six reasons why you might perform categorical encoding. These reasons are as follows:

- ML algorithms require numerical input
- Model interpretation and decision-making
- Handling of high cardinality
- Capturing the relationship between features and target
- Avoiding bias in model predictions
- Improving model performance

Next, you will learn about methods to perform categorical encoding, starting with one-hot encoding.

One-hot encoding

One-hot encoding is a simple yet powerful method for transforming categorical variables into binary vectors. For each category, this technique creates a binary column indicating the presence (1) or absence (0) of that category. The challenge with this method is that, as mentioned previously, if you have hundreds or thousands of unique categories, this method creates hundreds or thousands of new columns. This can make the model inefficient and can also lead to overfitting. One-hot encoding works best for cases where the number of categories is on the order of 10 unique values. For example, a Gender variable with categories female and male can be one-hot encoded into two columns: female (1 if female, 0 otherwise) and male (1 if male, 0 otherwise). Similarly, a Color variable with categories red, blue, and green would be encoded into three binary columns.

To reduce redundancy, (k-1) columns are often used for a variable with k categories. This approach is particularly useful for linear models where **multicollinearity** can be an issue. However, for tree-based models, encoding all k categories might be necessary.

Let's apply this encoding method to the two examples of categorical data we just mentioned:

1. To get started with this example, you will need to import pandas and the train_test_split and OneHotEncoder methods from scikit-learn:

    ```
    import pandas as pd
    from sklearn.model_selection import train_test_split
    from sklearn.preprocessing import OneHotEncoder
    ```

2. Next, you can set up some sample categorical data to use. This data has two columns one for Gender with values of male or female and one for Color with values of red, blue, and green. You can turn the sample data into a pandas DataFrame with df = pd.DataFrame(data):

    ```
    data = {
        'Gender': ['male', 'female', 'female', 'male'],
        'Color': ['red', 'blue', 'green', 'red']
    }
    df = pd.DataFrame(data)
    ```

3. Next, split the data using the train_test_split module. As discussed in previous chapters, you always want to split the dataset prior to doing any feature engineering work to prevent inflation of the model's accuracy:

    ```
    X_train, X_test = train_test_split(df, test_size=0.5,
        random_state=0)
    ```

4. Both pandas and scikit-learn offer one-hot encoding methods. Let's see how they compare. First, you can use pandas to do the encoding with pd.get_dummies. You can use drop_first=True to get *k*-1 columns out of *k* categories. This helps reduce redundancy. Print the output so you can see how pandas differs from scikit-learn:

```
pd_encoded = pd.get_dummies(X_train, drop_first=True)
print("Pandas One Hot Encoding:\n", pd_encoded)
```

5. Now, you can repeat the encoding with scikit-learn. pandas is more straightforward compared with scikit-learn for this type of encoding. With scikit-learn, you pass drop='first' to get the *k*-1 categories with this line: encoder = OneHotEncoder(drop='first', sparse_output=False). The sparse_output switch allows you to make a **compressed sparse row (CSR)** format matrix, which you don't need here. With scikit-learn's OneHotEncoder, you need to fit the encoder and then use the encoder to transform the data. You can do this with encoder.fit(X_train) and sk_encoded = encoder.transform(X_train). Finally, you can put the results into a pandas DataFrame and print the result:

```
encoder = OneHotEncoder(drop='first',
    sparse_output=False)
encoder.fit(X_train)
sk_encoded = encoder.transform(X_train)
sk_encoded_df = pd.DataFrame(sk_encoded,
    columns=encoder.get_feature_names_out())
print("Scikit-learn One Hot Encoding:\n",
    sk_encoded_df)
```

Here is what the X_train dataset looks like:

```
    Gender  Color
1   female  blue
0     male  red
```

6. Running the encoding results in the following output:

```
Pandas One Hot Encoding:
      Gender_male   Color_red
1               0           0
0               1           1
Scikit-learn One Hot Encoding:
      Gender_male  Color_red
0             0.0        0.0
1             1.0        1.0
```

Since there are only two choices in each column of the train dataset, the results of the *k*-1 one hot encoding are one column each (*k*=2, so *k*-1 is one). The value in the column indicates either male or not or red or not. In this case, you can guess the other option for gender might be female; however, you don't know what the other color option might be. For this reason, it is always a good idea to keep the original column when doing one-hot encoding so you don't lose information about the dataset.

The main difference between the pandas encoder and the scikit-learn encoder is how binary values are returned. pandas gives integer values, scikit-learn uses floating point values, pandas and scikit-learn use 0 for `False` and 1 for `True`. You may find uses for both methods; for example, if you are doing a lot of manipulation of the data using pandas, it is handy to use the pandas encoder. Similarly, if you will be using scikit-learn to model the data, using the scikit-learn encoder can be combined with the model in a pipeline. We will discuss pipelines in *Chapter 12*.

Now that you've applied one-hot encoding, what happens if there are a lot of categories and you do one-hot encoding? The situation of having a lot of categories is known as high cardinality. You will learn how to manage that in the next section.

One-hot encoding with frequent categories

High cardinality and rare labels in categorical variables can lead to overfitting or difficulties in scoring unseen categories. One-hot encoding of the most frequent categories mitigates these issues by creating binary variables for only the top categories, effectively grouping the less-common categories into a single category.

This method prevents the feature space from expanding excessively and is especially useful in situations where a few categories dominate. However, one-hot encoding with frequent categories may not capture all the information of less common labels.

Let's apply this to an example of housing prices. Let's get started:

1. You will be using `pandas`, as well as `train_test_split` and `OneHotEncoder` from scikit-learn, so you can start by importing those packages:

    ```
    import pandas as pd
    from sklearn.model_selection import train_test_split
    from sklearn.preprocessing import OneHotEncoder
    ```

2. You can set up some sample data to see how this method works. In this case, you can use a list of neighborhoods (the input variable) and house sale prices (the target variable), turn them into a pandas DataFrame, and split them into train and test datasets:

    ```
    data = {    'Neighborhood': ['OldTown', 'CollgCr',
        'Somerst', 'NAmes', 'OldTown', 'CollgCr',
        'Edwards', 'NAmes', 'Edwards', 'OldTown'],
        'SalePrice': [200000, 250000, 190000, 300000,
            210000, 230000, 180000, 320000, 190000,
            215000]
    ```

```
    }
df = pd.DataFrame(data)
print (df)
X_train, X_test, y_train, y_test = train_test_split(
    df[['Neighborhood']], df['SalePrice'],
    test_size=0.5, random_state=0)
print(X_train)
```

This example is not an extreme one; there are only five categories and you could use standard one-hot encoding without creating a huge number of additional columns in the dataset. Imagine, however, that you have a parameter that has hundreds of values, maybe a parameter for country names. If you use one-hot encoding, you get a column for each option, which can be a lot of columns. This method reduces that down to the most frequent values and buckets the infrequent ones into a single category.

Here's the sample dataset, df:

	Neighborhood	SalePrice
0	OldTown	200000
1	CollgCr	250000
2	Somerst	190000
3	NAmes	300000
4	OldTown	210000
5	CollgCr	230000
6	Edwards	180000
7	NAmes	320000
8	Edwards	190000
9	OldTown	215000

And here's X_train:

	Neighborhood
6	Edwards
7	NAmes
3	NAmes
0	OldTown
5	CollgCr

3. Next, you can create a function that finds the most common categories. To do this, you can sort the column you want to analyze in descending order by count of the occurrence of a value:

```
df[variable].value_counts().sort_values(ascending=False)
```

You can use a variable to set a value for how many different columns you want to create in the result with the how_many variable. Here, how_many is set to 2, which will produce two additional columns for the top two neighborhoods in terms of frequency of appearance in the Neighborhood column of the train dataset. The result of this function is a list that contains the top categories. You can apply it by passing `X_train` and `Neighborhood` to `calculate_top_categories`:

```
def calculate_top_categories(df, variable,
    how_many=2):
    return [x for x in df[
        variable].value_counts().sort_values(
            ascending=False).head(how_many).index]
top_categories = calculate_top_categories(X_train,
    'Neighborhood')
```

4. To use pandas to do the encoding with `top_categories`, set up a function that takes the training dataset, the test dataset, the column to encode, and the list of top variables as arguments. Then, in the function, loop through each of the labels in the `top_categories` list and make a new column for that value in both the train and test DataFrames. Use the function by calling `one_hot_encode` with `X_train`, `X_test`, `Neighborhood`, and `top_categories`. Print out `X_train_encoded` so you can compare the result from pandas encoding with scikit-learn's `OneHotEncoder`:

```
def one_hot_encode(train, test, variable,
    top_x_labels):
        for label in top_x_labels: train[
            variable + '_' + label] = (train[
                variable] == label).astype(int)
            test[variable + '_' + label] = (test[
                variable] == label).astype(int)

        return train, test

X_train_encoded, X_test_encoded = one_hot_encode(
    X_train, X_test, 'Neighborhood', top_categories)
print("Pandas One Hot Encoding of Frequent
    Categories:\n", X_train_encoded)
```

This gives the following result:

```
Pandas One Hot Encoding of Frequent Categories:
  Neighborhood Neighborhood_NAmes Neighborhood_Edwards
6      Edwards                  0                    1
7        NAmes                  1                    0
3        NAmes                  1                    0
0      oldTown                  0                    0
5      CollgCr                  0                    0
```

5. Now, you can try scikit-learn's OneHotEncoder with categories set manually using the top_categories list. Set sparse_output=False because you don't need to create a CSR format matrix. Set handle_unknown = 'ignore' to ignore any values that are in the DataFrame but not in the list of categories you've generated. With scikit-learn's OneHotEncoder, you need to fit the encoder and then use the encoder to transform the data. We do this with encoder.fit(X_train[['Neighborhood']]) and sk_encoded = encoder.transform(X_train[['Neighborhood']]). Finally, you can put the results into a pandas DataFrame and print the result:

```
encoder = OneHotEncoder(categories=[top_categories],
    sparse_output=False, handle_unknown='ignore')
encoder.fit(X_train[['Neighborhood']])

sk_encoded = encoder.transform(
    X_train[['Neighborhood']])
sk_encoded_df = pd.DataFrame(sk_encoded,
    columns=encoder.get_feature_names_out(
        ['Neighborhood'])
    )
print("Scikit-learn One Hot Encoding of Frequent
    Categories:\n", sk_encoded_df)
```

The results of the scikit-learn encoding look as follows:

```
Scikit-learn One Hot Encoding of Frequent Categories:
    Neighborhood_NAmes  Neighborhood_Edwards
0                  0.0                   1.0
1                  1.0                   0.0
2                  1.0                   0.0
3                  0.0                   0.0
4                  0.0                   0.0
```

So far, you have used both pandas and scikit-learn and applied two types of one-hot encoding to turn a list of categories into separate binary columns. While there are slight formatting differences in how the output is presented between pandas and scikit-learn, the encoded values are exactly the same. Next, you will learn about integer encoding.

Integer encoding

Integer encoding replaces categories with unique integers from 0 to *n*-1, where *n* is the number of distinct categories. The benefit of this method is that it does not expand the feature space and is computationally efficient. However, it does not capture any inherent relationships between categories. Let's see how this works when implemented in code:

1. For this method, you will use pandas, as well as `train_test_split` and `LabelEncoder` from scikit-learn, so you can start by importing those packages. You will be able to compare the results from integer encoding with pandas and scikit-learn's `LabelEncoder`. You can start by importing the packages you will need:

    ```
    import pandas as pd
    from sklearn.model_selection import train_test_split
    from sklearn.preprocessing import LabelEncoder
    ```

2. Now, you can set up some sample data to use, which will be lists of values for `Neighborhood` and `SalePrice` placed into a pandas DataFrame:

    ```
    data = {
        'Neighborhood': ['OldTown', 'CollgCr', 'Somerst',
        'NAmes', 'OldTown', 'CollgCr', 'Edwards', 'NAmes',
        'Edwards', 'OldTown'],
        'SalePrice': [200000, 250000, 190000, 300000,
            210000, 230000, 180000, 320000, 190000,
            215000]
    }
    df = pd.DataFrame(data)
    ```

3. Next, perform the train-test split on the sample data:

    ```
    X_train, X_test, y_train, y_test = train_test_split(
        df[['Neighborhood']], df['SalePrice'],
        test_size=0.5, random_state=0)
    ```

4. To do integer encoding in pandas, you will use scikit-learn's `LabelEncoder`. This is a hybrid approach using both scikit-learn and pandas, unlike the one-hot encoding, which could be done completely by pandas alone. First, create a column called `Neighborhood_enc`, which uses `LabelEncoder` to assign values to each of the neighborhood names within the training dataset:

    ```
    le = LabelEncoder()
    X_train['Neighborhood_enc'] = le.fit_transform(
        X_train['Neighborhood'])
    ```

5. You can't know in advance whether there are values in the Neighborhood column that occur in the test dataset that are not in the train set, so this code sets the value of neighborhoods in the test dataset that aren't in the train dataset to -1. Now, -1 is a better choice here than 0 since pandas indexing starts from 0. This means zero is used for a category already. Using it again to indicate an unseen category in the test data would confuse the two categories and cause the model to give the wrong results. This line of code looks through the values in the Neighborhood column of the X_test dataset and converts them to the values in the encoder or, if the encoder doesn't have that category, fills in with -1:

```
X_test['Neighborhood_enc'] = X_test[
    'Neighborhood'].apply(lambda x:
    le.transform([x])[0] if x in le.classes_ else -1)
```

6. Print out the results to see what this code does with the data:

```
print("Pandas Integer Encoding:\n", X_train)
print("Pandas Integer Encoding Test Set:\n", X_test)
```

This displays the following output:

```
Pandas Integer Encoding:
   Neighborhood  Neighborhood_enc
6      Edwards                 1
7        NAmes                 2
3        NAmes                 2
0      OldTown                 3
5      CollgCr                 0
Pandas Integer Encoding Test Set:
   Neighborhood  Neighborhood_enc
2      Somerst                -1
8      Edwards                 1
4      OldTown                 3
9      OldTown                 3
1      CollgCr                 0
```

Notice how Somerst, which was not in the train dataset, is encoded as -1. Notice also that the CollgCr neighborhood is encoded as 0.

7. To use scikit-learn to perform integer encoding, you will need the defaultdict module, so we import it from the collections package and then use it to create the mapping to an encoding value with d = defaultdict(LabelEncoder):

```
from collections import defaultdict
# Create a dictionary to hold label encoders for each #column
d = defaultdict(LabelEncoder)
```

8. Next, you can apply the encoding to the train dataset and test dataset using `fit_transform`, handling the situation where a value appears in the test dataset that isn't in the encoder by filling in with `-1`. Print the results to see how the encoding looks:

```
X_train['Neighborhood_enc'] = d['Neighborhood'].fit_transform(
    X_train['Neighborhood'])
X_test['Neighborhood_enc'] = X_test[
    'Neighborhood'].apply(lambda x:
        d['Neighborhood'].transform([x])[0] if x in
            d['Neighborhood'].classes_ else -1)
print("Scikit-learn Integer Encoding:\n", X_train)
print("Scikit-learn Integer Encoding Test Set:\n",
    X_test)
```

This produces the following:

```
Scikit-learn Integer Encoding:
    Neighborhood  Neighborhood_enc
6       Edwards                  1
7        NAmes                   2
3        NAmes                   2
0      OldTown                   3
5      CollgCr                   0

Scikit-learn Integer Encoding Test Set:
    Neighborhood  Neighborhood_enc
2      Somerst                  -1
8      Edwards                   1
4      OldTown                   3
9      OldTown                   3
1      CollgCr                   0
```

Scikit-learn and pandas produce equivalent results. Which you choose to use is up to you. If you are doing a lot of work in scikit-learn, then staying in that package has benefits. If you find the pandas approach easier to understand, use that instead.

Integer encoding is a very basic method for turning categories into numbers that a model can utilize by assigning a number to each category. Integer encoding is simple to use and effective for non-linear models. Next, you will learn about count or frequency encoding, which adds a layer of meaning to the values used in encoding the category.

Count or frequency encoding

Count or frequency encoding replaces categorical values with their occurrence counts or relative frequencies within the dataset. This technique allows us to represent categories numerically without expanding the feature space, which is useful when dealing with high-cardinality categorical features.

For instance, consider a dataset containing information about different neighborhoods and house characteristics. If a particular neighborhood (e.g., `OldTown`) appears 10 times in the dataset, count encoding would replace `OldTown` with 10, whereas frequency encoding would replace it with the fraction of times it appears (e.g., 10 out of 100 observations would be `0.1`). This method effectively captures the representation of each label but can lose valuable information if different categories have the same count or frequency. For example, if both `blue` and `red` appear 10 times, they will be encoded identically.

Let's apply count and frequency encoding using a small dataset, focusing on the categorical columns `Neighborhood` and `Exterior1st`. This example will illustrate how to convert these categorical columns into numerical representations using count and frequency encoding:

1. To get started, import `pandas` and import `train_test_split` from scikit-learn. Set up sample data with `Neighborhood`, as well as the exterior information of `Exterior1st` and `SalePrice`, and put those lists into a pandas DataFrame with `df = pd.DataFrame(data)`:

    ```
    import pandas as pd
    from sklearn.model_selection import train_test_split

    # Sample data
    data = {'Neighborhood': ['OldTown', 'CollgCr',
        'Somerst', 'NAmes', 'OldTown', 'CollgCr',
        'Edwards', 'NAmes', 'Edwards', 'OldTown'],
        'Exterior1st': ['VinylSd', 'MetalSd', 'Wd Sdng',
        'HdBoard', 'BrkFace', 'VinylSd', 'HdBoard',
        'MetalSd', 'HdBoard', 'VinylSd'],
        'SalePrice': [200000, 250000, 190000, 300000,
            210000, 230000, 180000, 320000, 190000,
            215000]
    }
    df = pd.DataFrame(data)
    ```

2. Next, split the dataset into train and test, with `SalePrice` as the target `y` value:

    ```
    # Separate into training and testing sets
    X_train, X_test, y_train, y_test = train_test_split(
        df[['Neighborhood', 'Exterior1st']],
        df['SalePrice'], test_size=0.5, random_state=0)
    ```

3. You can do count encoding with `.value_counts()` and the `.map` capabilities in pandas. First, use `.value_counts()` to get the count of each unique category in `Neighborhood` within the training set, and then convert these counts to a dictionary using `.to_dict()`. This dictionary is called `count_map`, and it contains the counts of each `Neighborhood` category.

    Next, we apply this dictionary to the `Neighborhood` column in `X_train` using the `.map()` function, which replaces each value in the column with its corresponding count from `count_map`. This creates a new column called `Neighborhood_count` in `X_train`.

    For the test dataset (`X_test`), we also use the same mapping approach to create a `Neighborhood_count` column. However, there may be categories present in the test set that were not seen in the training set, leading to NaN values. To handle this, we use `.fillna(0)` to replace these missing values with 0:

    ```
    # Count encoding with pandas
    count_map = X_train[
        'Neighborhood'].value_counts().to_dict()
    X_train['Neighborhood_count'] = X_train[
        'Neighborhood'].map(count_map)
    X_test['Neighborhood_count'] = X_test[
        'Neighborhood'].map(count_map).fillna(0)

    print("Pandas Count Encoding:\n", X_train)
    ```

    This prints the following output:

    ```
    Pandas Count Encoding:
        Neighborhood Exterior1st  Neighborhood_count
    6       Edwards    HdBoard                     1
    7         NAmes    MetalSd                     2
    3         NAmes    HdBoard                     2
    0       OldTown    VinylSd                     1
    5       CollgCr    VinylSd                     1
    ```

    Three neighborhoods (`Edwards`, `OldTown`, `CollgCr`) only show up once in the train dataset, so they are all encoded as '1'. The `NAmes` neighborhood has two rows, so it is encoded as '2'.

4. Next, you can add a column that uses frequency to encode the `Exterior1st` values. Start by creating a dictionary based on `value_counts()` as before. To get the frequency, divide by the total number of items in `X_train`, which is calculated with `len(X_train)`:

    ```
    # Frequency encoding with pandas
    frequency_map = (X_train[
        'Exterior1st'].value_counts() / len(
            X_train)).to_dict()
    ```

5. Apply the `frequency_map` dictionary to `X_train['Exterior1st']` and `X_test['Exterior1st']` to encode those columns. Use `.fillna(0)` when encoding the `X_test` dataset to manage categories that are in the test dataset that were not in the train dataset. Finally, print out the result:

```
X_train['Exterior1st_freq'] = X_train[
    'Exterior1st'].map(frequency_map)
X_test['Exterior1st_freq'] = X_test[
    'Exterior1st'].map(frequency_map).fillna(0)

print("Pandas Frequency Encoding:\n", X_train)
```

This outputs the following:

```
   Neighborhood Exterior1st  Neighborhood_count  Exterior1st_freq
6       Hdboard                               1               0.4
7         Names     MetalSd                   2               0.2
3         Names     HdBoard                   2               0.4
0       OldTown     VinylSd                   1               0.4
5       CollgCr     VinylSd                   1               0.4
```

The encoding for the `Exterior1st` column in `Exterior1st_freq` works like this. `HdBoard` and `VinylSd` both appear twice out of five values for a frequency of 2/5 = 0.4. `MetalSd` appears once out of five for a frequency of 1/5 = 0.2.

Now you've learned a simple encoding method that supplies some meaning to the value that replaces a word. Let's continue with that by learning about ordered integer encoding.

Ordered integer encoding

Ordered integer encoding replaces categories with integers according to the order of a target variable. This is often the mean of the target variable within each category. If we encode the `Neighborhood` column using this method, we will replace the neighborhood name with the average of the `SalePrice` column for that neighborhood. The ordered integer encoding method ensures a monotonic relationship between the encoded variable and the target, beneficial for linear models. Let's apply it using pandas:

1. First, import `pandas`, and `train_test_split` from scikit-learn to prepare for applying the encoding. Import `matplotlib.pyplot`, which you will use to make graphs to show the relationship between the encoded categorical variables and the target variable:

```
import pandas as pd
from sklearn.model_selection import train_test_split
import matplotlib.pyplot as plt
```

2. Next, set up the sample data and split it into train and test datasets using `train_test_split`. This sample dataset has two categorical columns, `Neighborhood` and `Exterior1st`, and a target column, `SalePrice`:

```
data = {'Neighborhood': ['OldTown', 'CollgCr',
    'Somerst', 'NAmes', 'OldTown', 'CollgCr',
    'Edwards', 'NAmes', 'Edwards', 'OldTown'],
    'Exterior1st': ['VinylSd', 'MetalSd', 'Wd Sdng',
        'HdBoard', 'BrkFace', 'VinylSd', 'HdBoard',
        'MetalSd', 'HdBoard', 'VinylSd'],
        'SalePrice': [200000, 250000, 190000, 300000,
            210000, 230000, 180000, 320000, 190000,
            215000]
}
df = pd.DataFrame(data)

# Separate into training and testing sets
X_train, X_test = train_test_split(df, test_size=0.5,
    random_state=0)
```

3. To perform the encoding, you need some way to order the values to associate the target values to each category. If the column you are encoding is ordinal, you can use that inherent order for the encoding. Say, for example, you have a column that has the values good, bad, and excellent. You can create a dictionary that equates bad to 0, good to 1, and excellent to 2 and use that for the encoding.

    In the sample data, there is no such ordering, so you can first calculate the mean of the target value for each category and use that to order the categories. The encoding is the number order of the ranked categories. The average `SalePrice` value for `VinylSd` is 215000. For HdBoard it is 240000, and for `MetalSd` it is 320000. This then provides the encoding of `VinylSd` = 0, HdBoard = 1, and `MetalSd` = 2. You might wonder, why not just use the mean value for the group as the encoding? You can. We will cover that in the next section, *Mean encoding*. For each category, the `find_category_mappings` function calculates the mean target value using `df.groupby([variable])[target].mean()` and sorts the values from lowest to highest:

```
# Function to calculate mean target value for each category and
create mapping
def find_category_mappings(df, variable, target):
    ordered_labels = df.groupby(
        [variable])[target].mean().sort_values().index
    return {k: i for i, k in enumerate(
        ordered_labels, 0)}
```

4. Next, you can create a function to apply the encoding to the train and test datasets:

```
# Function to apply the mapping
def integer_encode(train, test, variable,
    ordinal_mapping):
        train[variable + '_enc'] = train[
            variable].map(ordinal_mapping)
        test[variable + '_enc'] = test[
            variable].map(ordinal_mapping)
    return train, test
```

5. Use the functions you just created to apply the encoding. First, create the mappings for the Neighborhood column and apply them to X_train and X_test:

```
mappings = find_category_mappings(X_train,
    'Neighborhood', 'SalePrice')
X_train, X_test = integer_encode(X_train, X_test,
    'Neighborhood', mappings)
```

6. Then, repeat the same process for the Exterior1st column, X_train, and X_test:

```
mappings = find_category_mappings(X_train,
    'Exterior1st', 'SalePrice')
X_train, X_test = integer_encode(X_train, X_test,
    'Exterior1st', mappings)
```

7. To wrap up this section, you can print the top of X_train to see how the encoding has worked:

```
print("Ordered Integer Encoding with Pandas:\n",
    X_train.head())
```

This results in the following:

```
Ordered Integer Encoding with Pandas:
   Neighborhood Exterior1st  SalePrice  Neighborhood_
enc   Exterior1st_enc
6  Edwards      Hdboard      180000           0
              1
6  NAmes        MetalSd      320000           3
              2
3  NAmes        Hdboard      300000           3
              1
0  OldTown      VinylSd      200000           1
              0
5  CollgCr      VinylSd      230000           2
              0
```

8. Finally, you can visualize the relationship between the encoded columns and the target column SalePrice using Matplotlib; create two plots, one each for Neighborhood_enc and Exterior1st_enc, which plot the mean SalePrice for each category value:

```
# Let's visualize the monotonic relationship
for var in ['Neighborhood_enc', 'Exterior1st_enc']:
```

```
fig = plt.figure()
X_train.groupby([var])['SalePrice'].mean().plot()
plt.title(f'Monotonic relationship between {
    var} and SalePrice')
plt.ylabel('Mean SalePrice')
plt.show()
```

This outputs the following two graphs:

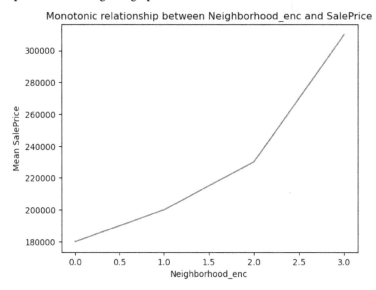

Figure 8.1 – Relationship between Neighborhood_enc and SalePrice

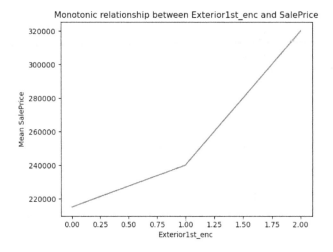

Figure 8.2 – Relationship between Exterior1st_enc and SalePrice

Now you've tried out ordered integer encoding on sample data and encoded two columns: `Neighborhood` and `Exterior1st`. Next, you will look at mean encoding, which makes a stronger connection between the encoding value and the target variable.

Mean encoding

Mean encoding transforms categorical variables into numerical values by replacing each category with the mean of the target variable for that category. This technique establishes a direct relationship between the encoded variable and the target, making it particularly useful in linear models. Let's implement it:

1. First, import `pandas` and `train_test_split` from scikit-learn to prepare for applying the encoding. Import `matplotlib.pyplot`, which you will use to make graphs showing the relationship between the encoded categorical variables and the target variable:

    ```
    import pandas as pd
    from sklearn.model_selection import train_test_split
    import matplotlib.pyplot as plt
    ```

2. Next, set up the sample data and split it into train and test datasets using `train_test_split`. This sample dataset has two categorical columns, `Neighborhood` and `Exterior1st`, and a target column, `SalePrice`:

    ```
    # Sample data
    data = {
        'Neighborhood': ['OldTown', 'CollgCr', 'Somerst',
        'NAmes', 'OldTown', 'CollgCr', 'Edwards', 'NAmes',
        'Edwards', 'OldTown'],
        'Exterior1st': ['VinylSd', 'MetalSd', 'Wd Sdng',
            'HdBoard', 'BrkFace', 'VinylSd', 'HdBoard',
            'MetalSd', 'HdBoard', 'VinylSd'],
        'SalePrice': [200000, 250000, 190000, 300000,
            210000, 230000, 180000, 320000, 190000,
            215000]
    }
    df = pd.DataFrame(data)

    # Separate into training and testing sets
    X_train, X_test, y_train, y_test = train_test_split(
        df, df['SalePrice'], test_size=0.5,
        random_state=0)
    ```

3. To perform the encoding, you will calculate the mean of the target value for each category and turn it into a dictionary:

```
# Function to calculate mean target value for each category and
create mapping
def find_category_mappings(df, variable, target):
    return df.groupby(
        [variable])[target].mean().to_dict()
```

4. Now, you are ready to apply this mapping. The mean_encode function replaces the text name of the category with the mean SalePrice value for that category:

```
# Function to apply the mapping
def mean_encode(train, test, variable,
    ordinal_mapping):
        train[variable + '_mean_enc'] = train[
            variable].map(ordinal_mapping)
    test[variable + '_mean_enc'] = test[
        variable].map(ordinal_mapping)
    return train, test
```

5. To encode the Neighborhood column, call find_category_mappings with the X_train dataset, the column to encode, Neighborhood, and which column has the target variables, SalePrice. Then, apply the encoding to X_train and X_test for the Neighborhood column:

```
# Apply mean encoding
mappings = find_category_mappings(X_train,
    'Neighborhood', 'SalePrice')
X_train, X_test = mean_encode(X_train, X_test,
    'Neighborhood', mappings)
```

6. To encode the Exterior1st column, call find_category_mappings with the X_train dataset, the column to encode, Exterior1st, and which column has the target variables, SalePrice. Then, apply the encoding to X_train and X_test for the Exterior1st column:

```
mappings = find_category_mappings(X_train,
    'Exterior1st', 'SalePrice')
X_train, X_test = mean_encode(X_train, X_test,
    'Exterior1st', mappings)
```

7. To finish this section, print out the encoded table:

```
print("Mean Encoding with Pandas:\n", X_train.head())
```

This produces the output:

```
Mean Encoding with Pandas:
   Neighborhood Exterior1st  SalePrice  Neighborhood_mean_enc  \
6       Edwards    HdBoard      180000                180000.0
7         NAmes    MetalSd      320000                310000.0
3         NAmes    HdBoard      300000                310000.0
0       OldTown    VinylSd      200000                200000.0
5       CollgCr    VinylSd      230000                230000.0

   Exterior1st_mean_enc
6             240000.0
7             320000.0
3             240000.0
0             215000.0
5             215000.0
```

8. You can plot the relationship between the target variable and the mean encoded columns, using Matplotlib to create two plots, one each for Neighborhood_enc and Exterior1st_enc, which plot the mean SalePrice for each category value:

```
# Let's visualize the monotonic relationship
for var in ['Neighborhood_mean_enc',
    'Exterior1st_mean_enc']:
        fig = plt.figure()
        X_train.groupby([var])[
            'SalePrice'].mean().plot()
        plt.title(f'Monotonic relationship between {
            var} and SalePrice')
        plt.ylabel('Mean SalePrice')
        plt.show()
```

This will display the following graphs:

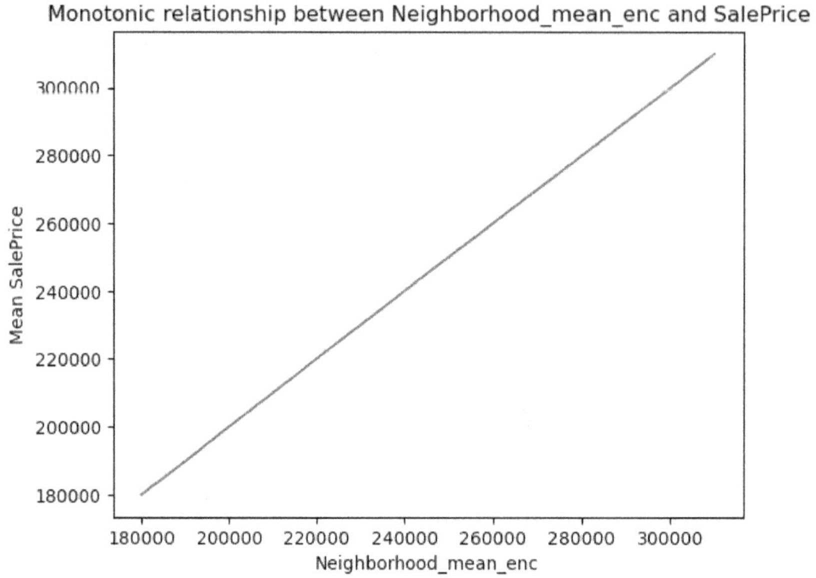

Figure 8.3 – Relationship between mean encoded Neighborhood and mean SalePrice

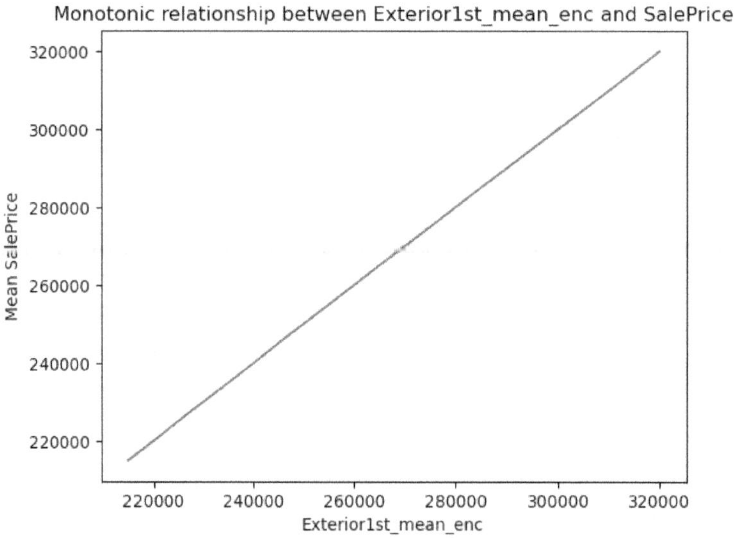

Figure 8.4 – Relationship between mean encoded Exterior1st and mean SalePrice

Now you've implemented mean encoding on the sample data. Next, you will learn about probability ratio encoding, which also uses the target variable to inform how to encode a categorical value.

Probability ratio encoding

Probability ratio encoding transforms categorical variables into numerical values by calculating the ratio of the probability of the target being 1 to the probability of it being 0 for each category. This encoding technique captures the distinct contribution of each category to the target outcome, providing valuable information for building effective models. Let's get started:

1. To begin, import `pandas` and `train_test_split` from scikit-learn to prepare for applying the encoding. Import `matplotlib.pyplot`, which you will use to make graphs showing the relationship between the encoded categorical variables and the target variable:

    ```
    import pandas as pd
    from sklearn.model_selection import train_test_split
    import matplotlib.pyplot as plt
    ```

2. Next, set up the sample data. This sample dataset has two categorical columns, `Neighborhood` and `Exterior1st`, and a target column, `SalePrice`:

    ```
    # Sample data
    data = {
        'Neighborhood': ['OldTown', 'CollgCr', 'Somerst',
        'NAmes', 'OldTown', 'CollgCr', 'Edwards', 'NAmes',
        'Edwards', 'OldTown'],
        'Exterior1st': ['VinylSd', 'MetalSd', 'Wd Sdng',
            'HdBoard', 'BrkFace', 'VinylSd', 'HdBoard',
            'MetalSd', 'HdBoard', 'VinylSd'],
        'SalePrice': [200000, 250000, 190000, 300000,
            210000, 230000, 180000, 320000, 190000,
            215000]
    }
    df = pd.DataFrame(data)
    ```

3. Probability ratio encoding works based on the target variable being either 0 or 1. The target variable in the sample data is not binary, so you need to convert it. An easy way to do this conversion is to decide on a threshold value and assess the target variable against that value, bucketing it as either lower or higher. The mean of the `SalePrice` values is just over 220000, so you can use 220000 as a dividing point to turn `SalePrice` into binary values and put the result into a column called `HighPrice`:

    ```
    # Binarize target for the purpose of this demonstration
    df['HighPrice'] = (df[
        'SalePrice'] > 220000).astype(int)
    ```

4. Next, split the data into train and test datasets using `train_test_split`:

```
# Separate into training and testing sets
X_train, X_test = train_test_split(df, test_size=0.5,
    random_state=0)
```

5. Now you can create a function that calculates the probability ratio, using the mean of the target variable. For each label, you calculate the probability of the target for the category to be 1 and the probability of the target being 0, with the final result being the ratio of the two probabilities. The smoothing factor is there to avoid division by zero, ensuring that the probability ratio encoding works correctly and provides meaningful values for all categories:

```
# Function to calculate probability ratio for each category with
smoothing
def find_category_mappings(df, variable, target,
    smoothing=1):
        tmp = pd.DataFrame(df.groupby(
            [variable])[target].mean())
        tmp['non-target'] = 1 - tmp[target]
        tmp['ratio'] = (tmp[
            target] + smoothing) / (tmp[
                'non-target'] + smoothing)
        return tmp['ratio'].to_dict()
```

Here's what the `tmp` DataFrame looks like for `Neighborhood` and `Exterior1st`:

	HighPrice
Neighborhood	
CollgCr	1.0
Edwards	0.0
NAmes	1.0
OldTown	0.0

	HighPrice
Exterior1st	
HdBoard	0.5
MetalSd	1.0
VinylSd	0.5

6. Create a function called `ratio_encode` to do the mapping using `.map`:

```
# Function to apply the mapping
def ratio_encode(train, test, variable,
    ordinal_mapping):
        train[variable + '_ratio_enc'] = train[
            variable].map(ordinal_mapping)
        test[variable + '_ratio_enc'] = test[
```

```
                    variable].map(ordinal_mapping)
         return train, test
```

7. To encode the `Neighborhood` and `Exterior1st` columns, call `find_category_mappings` with the `X_train` dataset, the column to encode, and which column has the target variables, `HighPrice`. Then, apply `ratio_encode` to perform the encoding of `X_train` and `X_test` for both the `Neighborhood` and `Exterior1st` columns:

```
# Apply probability ratio encoding
mappings = find_category_mappings(X_train,
    'Neighborhood', 'HighPrice')
X_train, X_test = ratio_encode(X_train, X_test,
    'Neighborhood', mappings)

mappings = find_category_mappings(X_train,
    'Exterior1st', 'HighPrice')
X_train, X_test = ratio_encode(X_train, X_test,
    'Exterior1st', mappings)
```

8. Finish this section by printing out the encoded table:

```
print("Probability Ratio Encoding with Pandas:\n",
    X_train.head())
```

This generates the following output:

```
Probability Ratio Encoding with Pandas:
    Neighborhood Exterior1st  SalePrice  HighPrice  Neighborhood_
ratio_enc  \
6       Edwards      HdBoard     180000          0        0.5
7        NAmes      MetalSd     320000          1        2.0
3        NAmes      HdBoard     300000          1        2.0
0      OldTown      VinylSd     200000          0        0.5
5       CollgCr      VinylSd     230000          1        2.0


    Exterior1st_ratio_enc
6                     1.0
7                     2.0
3                     1.0
0                     1.0
5                     1.0
```

9. You can plot the relationship between the `HighPrice` variable and the mean encoded columns, using Matplotlib to create two plots, one each for `Neighborhood_enc` and for `Exterior1st_enc`, which plot the mean `HighPrice` value for each category value:

```
# Let's visualize the monotonic relationship
for var in ['Neighborhood_ratio_enc',
    'Exterior1st_ratio_enc']:
        plt.figure()
        X_train.groupby([var])[
            'HighPrice'].mean().plot()
        plt.title(f'Monotonic relationship between {
            var} and HighPrice')
        plt.ylabel('Mean HighPrice')
        plt.xlabel(var)
        plt.show()
```

This will display the following output:

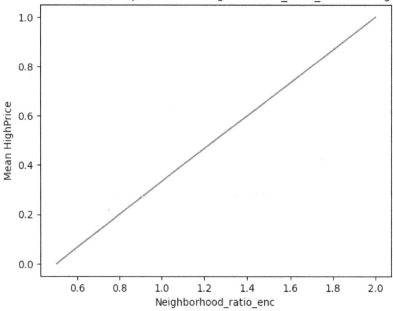

Figure 8.5 – Relationship between probability-encoded neighborhood and HighPrice

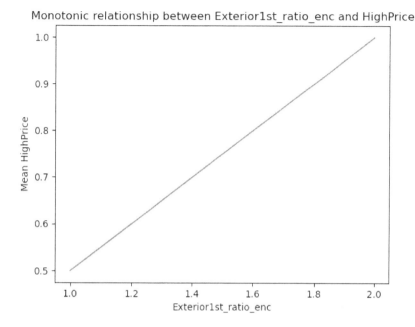

Figure 8.6 – Relationship between probability-encoded Exterior1st and HighPrice

Now you've applied probability ratio encoding. This method and mean encoding are both helpful for linear models in particular. Next, you will learn about WoE encoding, which has benefits for logistic regression.

Weight of Evidence (WoE) encoding

WoE encoding was originally developed for the credit and financial industries. It is used to build predictive models for assessing loan default risk in those industries. WoE quantifies the effectiveness of a grouping method by distinguishing between good and bad risks. This technique is particularly well suited for logistic regression as it establishes a monotonic relationship between the target variable and the independent variables, ordering the categories on a logistic scale. In this section, you can try it on sample data:

1. Start by importing `pandas` and importing `train_test_split` from scikit-learn to prepare for applying the encoding. Import `matplotlib.pyplot` and `numpy`, which you will use to make graphs to show the relationship between the encoded categorical variables and the target variable:

```
import pandas as pd
from sklearn.model_selection import train_test_split
import matplotlib.pyplot as plt
import numpy as np
```

2. Set up the sample data. This sample dataset has two categorical columns, `Neighborhood` and `Exterior1st`, and a target column, `SalePrice`:

```
# Sample data
data = {
    'Neighborhood': ['OldTown', 'CollgCr', 'Somerst',
    'NAmes', 'OldTown', 'CollgCr', 'Edwards', 'NAmes',
    'Edwards', 'OldTown'],
    'Exterior1st': ['VinylSd', 'MetalSd', 'Wd Sdng',
        'HdBoard', 'BrkFace', 'VinylSd', 'HdBoard',
        'MetalSd', 'HdBoard', 'VinylSd'],
    'SalePrice': [200000, 250000, 190000, 300000,
        210000, 230000, 180000, 320000, 190000,
        215000]
}
df = pd.DataFrame(data)
```

3. WoE encoding is used with an assessment of a target being either good or bad. For the purpose of this example, let's say `SalePrice` greater than `220000` is good, and other values are bad. You can create a column, `HighPrice`, set to `1` if `SalePrice` > `220000` and `0` otherwise with this code:

```
# Binarize target for the purpose of this demonstration
df['HighPrice'] = (df[
    'SalePrice'] > 220000).astype(int)
```

4. Next, split the data into train and test datasets using `train_test_split`:

```
# Separate into training and testing sets
X_train, X_test = train_test_split(df, test_size=0.5,
    random_state=0)
```

5. WoE measures the strength of grouping, which separates good from bad. The equation for this is as follows:

$$WoE = \left[\ln\left(\frac{Dist\ Good}{Dist\ Bad} \right) \right] \times 100$$

This `calculate_woe` function uses NumPy's `np.log` to implement this equation and calculate the WoE for `target` for each value of `variable`:

```
# Function to calculate WoE for each category with smoothing
def calculate_woe(df, variable, target,
    smoothing=0.5):
    tmp = pd.DataFrame(df.groupby(
        [variable])[target].mean())
    tmp['non-target'] = 1 - tmp[target]
```

```
    tmp['woe'] = np.log((tmp[target] + smoothing) / (
        tmp['non-target'] + smoothing))
    return tmp['woe'].to_dict()
```

6. Create a `woe_encode` function to do the mapping using `.map`:

```
# Function to apply WoE mapping
def woe_encode(train, test, variable, woe_mapping):
    train[variable + '_woe_enc'] = train[
        variable].map(woe_mapping)
    test[variable + '_woe_enc'] = test[
        variable].map(woe_mapping)
    return train, test
```

7. To encode the `Neighborhood` column, call `calculate_woe` with the `X_train` dataset, the column to encode, and which column has the target variables, `HighPrice`. Then, apply `woe_encode` to perform the encoding of `X_train` and `X_test` for the `Neighborhood` column. The last line, `woe_mappings`, displays the content of the encoding so you can see it:

```
# Apply WoE encoding
woe_mappings = calculate_woe(X_train, 'Neighborhood',
    'HighPrice')
X_train, X_test = woe_encode(X_train, X_test,
    'Neighborhood', woe_mappings)
woe_mappings
```

This produces the following:

```
{'CollgCr': 1.0986122886681098,
 'Edwards': -1.0986122886681098,
 'NAmes': 1.0986122886681098,
 'OldTown': -1.0986122886681098}
```

In this section, you used pandas to apply WoE encoding. This method replaces categories with the log of the odds ratio of the target variable within each category, ensuring a monotonic relationship that is beneficial for logistic regression models. Next, you will learn how to deal with rare categories.

Engineering rare categories

Rare categories in categorical variables are non-numeric values that occur infrequently within a dataset. These seldom-seen categories can pose challenges during model training and scoring, potentially leading to overfitting or being absent in the test set. To address this, rare categories can be consolidated into a new category, such as `Rare` or `Other`. This is the same method you used in *Chapter 7* applied to the `Neighborhood` column and the `Exterior1st` column in this dataset. Let's put this into practice:

1. To begin, import `pandas`, and from scikit-learn, import `train_test_split`. Import `matplotlib.pyplot` to make graphs showing the relationship between the encoded categorical variables and the target variable:

```
import pandas as pd
from sklearn.model_selection import train_test_split
import matplotlib.pyplot as plt
```

2. Set up the sample data. This sample dataset has three categorical columns, `Neighborhood`, `Exterior1st`, and `Exterior2nd`, as well as a target column, `SalePrice`. Put the data into a pandas DataFrame with `df=pd.DataFrame(data)`:

```
# Sample data
data = {
    'Neighborhood': ['OldTown', 'CollgCr', 'Somerst',
    'NAmes', 'OldTown', 'CollgCr', 'Edwards', 'NAmes',
    'Edwards', 'OldTown'],
    'Exterior1st': ['VinylSd', 'MetalSd', 'Wd Sdng',
        'HdBoard', 'BrkFace', 'VinylSd', 'HdBoard',
        'MetalSd', 'HdBoard', 'VinylSd'],
    'Exterior2nd': ['VinylSd', 'MetalSd', 'Wd Sdng',
        'HdBoard', 'BrkFace', 'VinylSd', 'HdBoard',
        'MetalSd', 'HdBoard', 'VinylSd'],
    'SalePrice': [200000, 250000, 190000, 300000,
        210000, 230000, 180000, 320000, 190000,
        215000]
}
df = pd.DataFrame(data)
```

3. Next, split the data into train and test datasets using `train_test_split`. You can print the original dataset to see how it looks before encoding rare categories:

```
# Separate into training and testing sets
X_train, X_test, y_train, y_test = train_test_split(
    df.drop(columns=['SalePrice']), df['SalePrice'],
        test_size=0.5, random_state=0)
print("Original dataset:\n", X_train)
```

This results in the following output:

```
Original dataset:
   Neighborhood Exterior1st Exterior2nd
6       Edwards     HdBoard     HdBoard
7         NAmes     MetalSd     MetalSd
3         NAmes     HdBoard     HdBoard
0       OldTown     VinylSd     VinylSd
5       CollgCr     VinylSd     VinylSd
```

4. Now, create a function to replace rare categories with the word `'Rare'`. This function is designed so you can pass in the DataFrame, `df`, a list of columns to encode, `variables_list`, and a tolerance variable, `tolerance`. For each variable in `variables_list`, the normalized frequency of each value, `freq = df[variable].value_counts(normalize=True)` is compared to `tolerance`, and the results of the comparison are put into a list called `rare_categories`. Then a lambda function replaces the values with Rare if they are in the `rare_categories` list:

```python
def rare_encoding(df, variables_list, tolerance):
    for variable in variables_list:
        freq = df[variable].value_counts(
            normalize=True)
        rare_categories = freq[freq < tolerance].index
        df[variable] = df[variable].apply(lambda x:
            'Rare' if x in rare_categories else x)
    return df
```

5. To apply this encoding, call `rare_encoding` with the datasets to encode, `X_train` and `X_test`; a list of columns to encode, `variables`; and the tolerance to use, `0.3`. The tolerance is set by experimentation and seeing how much or how little data is placed into the Rare group. Print the dataset to see how the encoding looks:

```python
# Apply rare encoding to the dataset
variables = ['Neighborhood', 'Exterior1st',
    'Exterior2nd']
X_train_enc = rare_encoding(X_train.copy(), variables,
    0.3)
X_test_enc = rare_encoding(X_test.copy(), variables,
    0.3)
print("Engineering Rare Categories with Pandas:\n",
    X_train_enc.head())
```

This produces the following result:

```
Engineering Rare Categories with Pandas:
   Neighborhood Exterior1st Exterior2nd
6          Rare     HdBoard     HdBoard
7         NAmes        Rare        Rare
3         NAmes     HdBoard     HdBoard
0          Rare     VinylSd     VinylSd
5          Rare     VinylSd     VinylSd
```

6. To finish this section, you can visualize the distribution of categories before and after rare encoding. To do this, you can loop over the list of categorical columns and make bar graphs showing `value_counts` for each category before and after encoding:

```
for col in ['Neighborhood', 'Exterior1st',
    'Exterior2nd']:
        fig, ax = plt.subplots(1, 2, figsize=(10, 5))
    X_train[col].value_counts(
        normalize=True).sort_values().plot.bar(
            ax=ax[0], title=f'{col} before encoding')
    X_train_enc[col].value_counts(
        normalize=True).sort_values().plot.bar(
            ax=ax[1], title=f'{col} after encoding')
    ax[0].axhline(y=0.3, color='red')
    ax[1].axhline(y=0.3, color='red')
    plt.show()
```

This makes the following graphs:

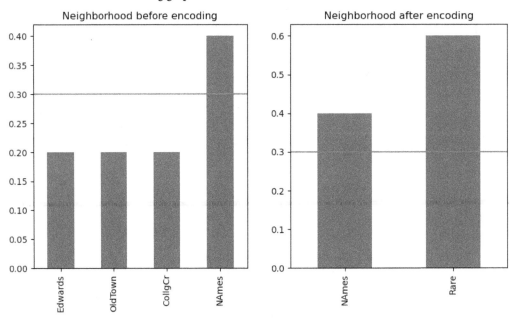

Figure 8.7 – Before and after rare encoding for neighborhood

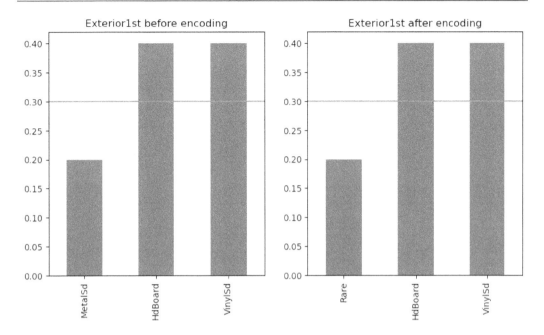

Figure 8.8 – Before and after rare encoding for exterior1st

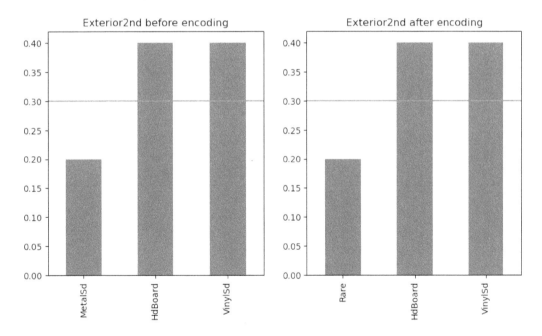

Figure 8.9 – Before and after rare encoding for exterior2nd

For the `Neighborhood` column, rare encoding results in combining a few of the categories, which affect counts for each category. For the other two columns, there is no combination happening, so the encoding results in renaming one category to `Rare`.

In this section, you've learned how to manage rare categories when encoding categorical variables. This concludes our list of methods for categorical encoding we wish to discuss. A list of all the encoding methods for categorical variables that are covered in this chapter, with their advantages and disadvantages, is given in the following table (*Table 8.1*):

Categorical Encoding Method	Advantages	Disadvantages
One-hot encoding	Simple method for transforming categorical variables into binary vectors.	Can expand the feature space excessively in the case of a high number of unique values in a categorical column.
One-hot encoding with frequent categories	Creates binary variables for only the top categories, effectively grouping less-common categories into a single category.	Can lose information due to grouping categories together.
Integer encoding	The benefit of this method is that it does not expand the feature space and is computationally efficient.	Does not capture any inherent relationships between categories.
Count or frequency encoding	Works particularly well when there are a lot of unique values in a categorical column.	Can lose information if different categories have the same count or frequency because it will group them together.
Ordered integer encoding	Creates a direct relationship between the encoded variable and the target, making it particularly useful in linear models.	May not suit your data or the type of model you wish to use.
Probability ratio encoding	Encodes the contribution of each category to the target outcome based on probabilities.	Requires the target variable to be converted into a binary, so this method may not be useful when the target is a continuous numeric variable.

Categorical Encoding Method	Advantages	Disadvantages
Weight of evidence (WoE)	Quantifies the effectiveness of a grouping method by distinguishing between good and bad risks. This technique is particularly well suited for logistic regression.	May not suit your data or the type of model you wish to use.
Rare categories	Combining rare categories into a single value of `Rare` or `Other` will help reduce overfitting or inflated accuracy due to examples being absent in the training or test dataset.	Loss of information due to clumping values together into a single category.

Table 8.1 – An overview of the encoding methods for categorical variables

Now, we're ready to summarize this chapter.

Summary

In this chapter, you explored various encoding techniques for transforming categorical variables into a format suitable for predictive modeling. From mean encoding, which establishes a direct relationship between the encoded variable and the target, to more specialized techniques such as probability ratio encoding and WoE encoding, each method offers unique advantages depending on the specific characteristics of the data and the requirements of the model. Additionally, you learned strategies for handling rare categories, ensuring that these infrequent occurrences do not negatively impact model performance. By carefully selecting and applying these encoding techniques, you can significantly enhance the predictive power and robustness of your models.

In the next chapter, *Chapter 9*, you will apply the data cleaning methods and techniques from *Chapter 6*, the feature selection techniques you learned in *Chapter 7*, and the encoding methods in this chapter while building models for time-series data using XGBoost.

Using XGBoost for Time Series Forecasting

As you learned in the preceding chapters, **XGBoost** is a powerful ensemble learning technique that builds a series of decision trees and aggregates their predictions to produce robust, accurate models. XGBoost excels in handling various regression and classification tasks by capturing complex, non-linear relationships within the data. This chapter focuses specifically on leveraging XGBoost for **time series forecasting**—also referred to as **time series prediction**, which means the same. Forecasting is a critical application in fields such as finance, supply chain management, and energy demand planning.

Time series forecasting presents unique challenges due to the sequential and temporal nature of the data. Unlike traditional supervised learning tasks where observations are independent and identically distributed, time series data points are intrinsically ordered and often exhibit trends, seasonality, and autocorrelation. XGBoost, though inherently designed for non-sequential data, can be adapted for time series forecasting by transforming the time-dependent data into a supervised learning format. This involves creating lag features, extracting meaningful components from the dates or timestamps associated with each observation, such as extracting the day of the week, month, quarter, or even whether a date falls on a weekend or holiday, and incorporating rolling statistics to enable XGBoost to effectively capture temporal dependencies.

In this chapter, you will gain exposure to the entire process of applying XGBoost for time series forecasting. You will learn about the following:

- Time series data characteristics and components
- Why use XGBoost for time series forecasting?
- Preparing time series data for XGBoost
- Practical considerations for handling time series data
- Training the XGBoost model
- Evaluating the model

- Forecasting future values

- Improving future predictions

By the end of this chapter, you will have a comprehensive understanding of how to harness the power of XGBoost for time series forecasting, building upon the foundational knowledge of XGBoost you gained in earlier chapters.

Technical requirements

To effectively follow along with the code examples and exercises in this chapter on time series modeling, you will need to have the following Python libraries. All code examples and datasets used in this chapter are available on our GitHub repository at `https://github.com/PacktPublishing/XGBoost-for-Regression-Predictive-Modeling-and-Time-Series-Analysis/blob/main/ch9/xgboost_timeseries_forecasting.ipynb`.

You will need the following Python packages:

- pandas 1.4.2

- NumPy 1.21.5

- XGBoost 1.7.3

- scikit-learn 1.2.2

- Matplotlib

Time series data – characteristics and components

There are many uses for data that include a time component, be it minutes, days, months, or seasons. Some uses are for forecasting demand for products that are influenced by seasonality, or for predicting maintenance schedules for manufacturing equipment. Data in the form of measurements taken at intervals indicated by a date or time value is known as time series data. This type of data can present trends where values are rising or falling over time, seasonality where there is a repeating pattern by the time of year, cyclical patterns, or even random unpredictable variation. Consider the following line plot (created by the author) as an example:

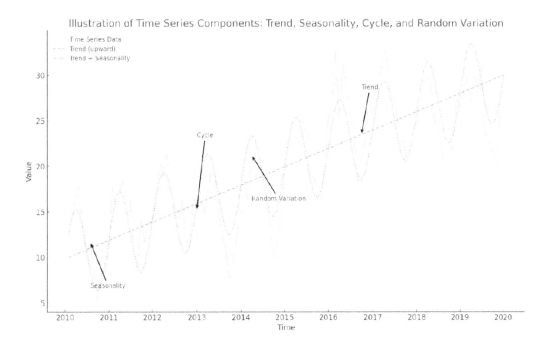

Figure 9.1 – A line plot with time on the x-axis (spanning multiple years) and
a sample metric (such as sales or temperature) on the y-axis

As demonstrated in the preceding figure, there are four key components to the time series:

- **Trend**: A gradual upward slope over several years, showing a consistent increase in value over time.

- **Seasonality**: Regular peaks and troughs within each year, such as higher values in the summer and lower in winter, capturing repeating seasonal patterns.

- **Cyclical pattern**: Longer, less regular fluctuations lasting more than a year, resembling economic cycles or market shifts.

- **Random variation**: Irregular small spikes and dips overlaying the plot, representing random, unpredictable changes that don't follow any specific pattern.

Analysis of this type of data and the ability to use historical data to forecast future values is used in finance, retail, manufacturing, and healthcare industries, to name a few. Next, we will discuss why XGBoost can be used for time series forecasting.

Why use XGBoost for time series forecasting?

In time series forecasting, selecting the right model is essential to capturing the underlying patterns and making accurate forecasts. Statistical time series models, such as **Autoregressive Integrated Moving Average** (**ARIMA**) and **Seasonal ARIMA** (**SARIMA**), and exponential smoothing methods are often effective for stable data with consistent trends and seasonality. However, they tend to perform poorly when the data exhibits frequent spikes or sudden changes. These models generally assume linear relationships and are designed to work best with stationary data—data without large fluctuations in mean or variance. When faced with high volatility or abrupt shifts, such models may struggle to capture the underlying structure accurately, leading to less reliable forecasts (Box, Jenkins, and Reinsel, 2015; Brockwell and Davis, 2016). In contrast, XGBoost's flexibility with non-linear relationships and robustness to outliers make it better suited for data with such irregularities. There are several advantages XGBoost offers that make it a compelling choice for time series forecasting:

- **Handling complex data relationships**: Traditional time series models, such as ARIMA and SARIMA, often assume linear relationships, which may limit their effectiveness with non-linear dependencies. In contrast, XGBoost, with its ensemble of decision trees, can capture complex, non-linear interactions between features. These features can include lagged values (e.g., previous sales figures or temperature readings) and external factors that may influence the target variable. Examples of these external factors include seasonal indicators (such as holidays or weekends), economic indicators (such as inflation rates or unemployment rates), and weather conditions (such as temperature or rainfall). By incorporating these variables, XGBoost can capture a broader range of influences, resulting in more accurate forecasts.

- **Feature engineering flexibility**: XGBoost thrives when provided with well-engineered features. For time series data, this includes lag features, rolling statistics, and date-based features. The flexibility in feature engineering allows XGBoost to model various aspects of the time series, such as trends and seasonality. Later in this chapter, we will create features that support time-series modeling.

- **Robustness to missing data and outliers**: XGBoost is inherently robust to missing data and can handle outliers effectively, which are common challenges in real-world time series datasets.

- **Scalability and performance**: XGBoost is optimized for speed and can handle large datasets efficiently, making it suitable for applications requiring rapid forecasting over extensive time horizons.

- **Integration of external variables**: XGBoost allows for the integration of **exogenous variables**—external factors that influence the target variable but aren't part of its past values. For example, in forecasting product demand, exogenous variables might include marketing spend or holidays, while weather forecasts could incorporate temperature or seasonal patterns. Including these variables helps the model capture a broader range of influences, enhancing forecast accuracy.

- **Handling non-stationary data**: As mentioned earlier in this chapter, traditional models, such as ARIMA and SARIMA, require the data to be stationary—meaning its mean, standard deviation, and other statistical properties do not significantly change over time—to make accurate predictions. In contrast, XGBoost does not impose such assumptions, allowing it to work more effectively with non-stationary data.

- **Parallel processing**: XGBoost supports parallel and distributed computing, which accelerates model training, especially with large datasets.

These advantages position XGBoost as a versatile and powerful tool for time series forecasting, capable of outperforming traditional models in various scenarios. Next, you'll learn about how to prepare time series data to be modeled using XGBoost.

Preparing time series data for XGBoost

Applying XGBoost to time series forecasting requires transforming the inherently sequential and temporal data into a supervised learning format that the model can process. This involves several key steps, including creating lag features, incorporating date-based and rolling statistical features, and ensuring that the temporal order is preserved when splitting the data into training and testing sets. Time series data is fundamentally different from regular tabular data because the order of observations matters. The goal of data preparation here is to provide the model with features that encode time-related information from the time series. This has the result of transforming the data into a supervised learning problem much like the housing value and housing price examples you explored in previous chapters. You'll apply this concept beginning with creating lag features. Let's do that next.

Creating lag features

One way to encode time-related information from time series data is to create input features that represent the values at previous time steps (lag features). For example, to predict the value at time *t*, you may use the values from *t-1*, *t-2*, and *t-3* as input features. This transformation allows XGBoost to learn from historical data. In this section, you will create lag features. Let's get started:

1. To begin, you will need pandas and NumPy to work with the data. For this example, you will use a synthetic time series dataset, which you can create by starting with the `date_range` function in pandas to create a `Date` column. You can specify the starting date with `start='1/1/2020'`, and how many items with `periods=100`. The `freq='D'` argument sets the difference between adjacent points to be in days. Then, add a `Value` column of a cumulative sum of 100 normally distributed random values using `np.random.randn(100).cumsum()`. End by setting the `Date` column to be the index for the DataFrame:

```
import pandas as pd
import numpy as np
```

```
# Create a synthetic time series dataset
date_range = pd.date_range(start='1/1/2020',
    periods=100, freq='D')
data = pd.DataFrame({'Date': date_range,
    'Value': np.random.randn(100).cumsum()})
data.set_index('Date', inplace=True)
```

The top few rows of this dataset will look like this:

```
              Value
Date
2020-01-01    1.222059
2020-01-02    2.764445
2020-01-03    4.330092
2020-01-04    6.509512
2020-01-05    7.937209
```

The dataset will look similar to this when plotted; it's randomly generated, so it will not be an exact match:

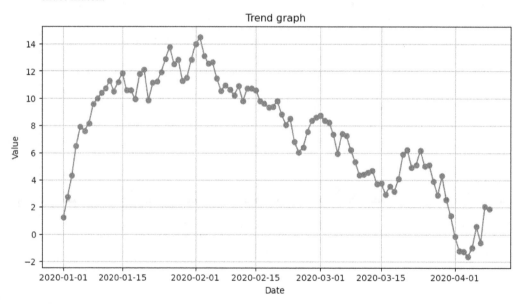

Figure 9.2 – Trend graph of the synthetic data generated for this example

2. Next, you can create a `create_lagged_features` function, which generates lagged values by creating new columns that contain previous data points for each time step. Here, we use three lags (i.e., `lags=3`), which creates three new columns: `lag_1`, `lag_2`, and `lag_3`. Each of these columns represents the value from 1, 2, and 3 days prior to the current date.

Let's walk through a few rows to see how these lagged values are populated:

- Row for 2020-01-04:

  - Value: 6.509512

  - lag_1: 4.330092 (value from 2020-01-03)

  - lag_2: 2.764445 (value from 2020-01-02)

  - lag_3: 1.222059 (value from 2020-01-01)

- Row for 2020-01-05:

  - Value: 7.937209

  - lag_1: 6.509512 (value from 2020-01-04)

  - lag_2: 4.330092 (value from 2020-01-03)

  - lag_3: 2.764445 (value from 2020-01-02)

- Row for 2020-01-06:

  - Value: 7.614653

  - lag_1: 7.937209 (value from 2020-01-05)

  - lag_2: 6.509512 (value from 2020-01-04)

  - lag_3: 4.330092 (value from 2020-01-03)

By using the .shift() method, each lag column references data from previous days, allowing the model to "look back" at prior values when making predictions. The function also uses df.dropna(inplace=True) to remove rows at the top of the dataset, where there aren't enough prior values to fill all lag columns (e.g., for the first row, we have no lagged values; for the second, only one lagged value is available). This setup ensures the dataset is complete and ready for modeling:

```
def create_lagged_features(data, lags=3):
    df = data.copy()
    for lag in range(1, lags + 1):
        df[f'lag_{lag}'] = df['Value'].shift(lag)
    df.dropna(inplace=True)
# Drop rows with NaN values generated by lagging
    return df
```

3. Create the new columns by calling the function and passing it the DataFrame, and decide how many lags to make:

```
lagged_data = create_lagged_features(data, lags=3)
print(lagged_data.head())
```

This outputs the following:

	Value	lag_1	lag_2	lag_3
Date				
2020-01-04	6.509512	4.330092	2.764445	1.222059
2020-01-05	7.937209	6.509512	4.330092	2.764445
2020-01-06	7.614653	7.937209	6.509512	4.330092
2020-01-07	8.151808	7.614653	7.937209	6.509512
2020-01-08	9.572572	8.151808	7.614653	7.937209

In this section, you generated lagged values for the time series of synthetic data. For each data point, you included the values from *t-1*, *t-2*, and *t-3* as input features in new columns.

In the next section, you will add features that are based on the date values.

Adding date-based features

Date-based features are an important aspect of time series data, and they can help capture cyclical or seasonal patterns. These features include attributes such as the day of the week, month, or even whether the date falls on a weekend or holiday. Adding these features can improve the model's understanding of the temporal context of the data. In this section, we will add date-based features to the synthetic data you've already generated. Let's begin:

1. You can build a function to add date-based features. In the synthetic dataset, the index consists of the date values. We'll use df.index.month to get the month for the date. We can also get the day of the week with df.index.dayofweek. Setting a flag for if a date is on a weekend is more involved – to do this, you need to know that each day of the week returns values from 0 to 6 with Monday = 0 and Sunday = 6. So, the weekends are given by the day of the week being >=5. You can use a lambda function to apply this logic:

```
def add_date_features(df):
    df['month'] = df.index.month
    df['day_of_week'] = df.index.dayofweek
    df['is_weekend'] = df['day_of_week'].apply(
        lambda x: 1 if x >= 5 else 0)
    return df
```

2. To add date features to the dataset, call the `add_date_features` function and pass in the `lagged_data` DataFrame. Then, you can print the top rows with `print(lagged_data.head())`:

```
lagged_data = add_date_features(lagged_data)
print(lagged_data.head())
```

This prints the following:

```
                Value       lag_1       lag_2       lag_3  \
Date
2020-01-04   6.509512    4.330092    2.764445    1.222059
2020-01-05   7.937209    6.509512    4.330092    2.764445
2020-01-06   7.614653    7.937209    6.509512    4.330092
2020-01-07   8.151808    7.614653    7.937209    6.509512
2020-01-08   9.572572    8.151808    7.614653    7.937209


                month   day_of_week   is_weekend
Date
2020-01-04        1              5            1
2020-01-05        1              6            1
2020-01-06        1              0            0
2020-01-07        1              1            0
2020-01-08        1              2            0
```

As you see, you have been able to generate new features. The month column encodes the month, 1-12, with 1 for January, 2 for February, and so on. The day of the week captures the day with values from 0 to 6 with Monday = 0 and Sunday = 6. Is_weekend is set to 1 (true) if the day of the week is Saturday = 5 or Sunday = 6, it is set to 0 (false) otherwise.

In this section, you created data-based features that can capture cyclical or seasonal patterns. Next, we will add statistics, such as a rolling mean, to comprehend trends.

Rolling statistics

Rolling statistics look at the stability of a time series by calculating statistics for values inside a sliding window, for example, looking at the weekly average for a column. This is calculated by taking the average of all the values in a column that occurred during a given week. Then, the window is moved forward to the next week and the average for that week is calculated. Rolling statistics can also be called moving statistics (e.g., moving average or a running average) and can be done over various size windows. **Rolling statistics** such as **moving averages** or standard deviations provide additional information about the trend or volatility in time series data. These statistics help the model capture short-term trends and variations, making it more effective at forecasting. By adding a **rolling mean** and **rolling standard deviation**, the model can better understand short-term trends and fluctuations in the data. Let's add these:

1. Begin by creating a function to calculate the rolling mean and rolling standard deviation. The pandas library has windowing support, which makes these calculations straightforward. You can call df['Value'].rolling and pass it the size of the window you'd like to use when calculating statistics. That looks like this:

```
def add_rolling_features(df, window=3):
    df['rolling_mean'] = df['Value'].rolling(
        window=window).mean().shift(1)
    df['rolling_std'] = df['Value'].rolling(
        window=window).std().shift(1)
    df.dropna(inplace=True)
    # Drop rows with NaN values
    return df
```

In this example, you've used a window size of 3; expanding the window size impacts the amount of smoothing performed by increasing the number of points included in the average and standard deviation. When calculating the rolling standard deviation, it's important to note that a minimum number of data points is required. For instance, with a window size of 3, the standard deviation can only be calculated once there are at least three data points in the window. Until then, the result is set as NaN. By shifting the calculations by one time step (using .shift(1)) and dropping NaN values, we ensure that each row's rolling features only include past data points, preserving the integrity of the time series structure.

2. Next, apply the function and print lagged_data.head() to see how these calculations look with a window size of 3:

```
lagged_data = add_rolling_features(lagged_data,
    window=3)
print(lagged_data.head())
```

The output is as follows:

```
              Value        lag_1       lag_2       lag_3\
Date
2020-01-07   8.151808     7.614653    7.937209    6.509512
2020-01-08   9.572572     8.151808    7.614653    7.937209
2020-01-09   9.980086     9.572572    8.151808    7.614653
2020-01-10  10.408591     9.980086    9.572572    8.151808
2020-01-11  10.718030    10.408591    9.980086    9.572572


              month    day_of_week    is_weekend \
Date
2020-01-07       1             1              0
```

2020-01-08	1	2	0
2020-01-09	1	3	0
2020-01-10	1	4	0
2020-01-11	1	5	1

	rolling_mean	rolling_std
Date		
2020-01-07	7.353791	0.748743
2020-01-08	7.901223	0.270380
2020-01-09	8.446344	1.011645
2020-01-10	9.234822	0.959795
2020-01-11	9.987083	0.418053

In this output, you see the original `value` column, followed by `lag_1`, which delays the contents of the `value` column by one row, `lag_2` which delays by two rows, and `lag_3`, which delays by three rows. Since each row is a day in this dataset, `lag_1` is the contents of the `value` column collected yesterday and is a one-day delay, `lag_2` is from the day before yesterday, and `lag_3` has the contents of the `value` column from three days ago. Then, we have our encoding columns, where we see each month encoded as a number with 1 = January, 2 = February, 3 = March, and so forth. The `day_of_week` column presents a numeric representation for Monday, Tuesday, Wednesday, and so on. The `is_weekend` column is a binary value, with 1 indicating that the date falls on a weekend and 0 indicating a weekday. Lastly, the `rolling_mean` column presents the mean of the last three values ((`lag_1` + `lag_2` + `lag_3`) / 3), and the final column, `rolling_std`, gives the rolling standard deviation, or spread, of the last three values: standard deviation (`lag_1`, `lag_2`, `lag_3`).

In this section, we created rolling statistics to look at trends and variations in the data. In the next section, we will split the time series data to get ready for model training.

Practical considerations for handling time series data

 When working with time series data, respecting the temporal order of observations is crucial. Unlike typical machine learning tasks, where data can be randomly shuffled, time series data must be split sequentially, so that the training set contains earlier observations, while the test set contains more recent ones. This ensures the model is evaluated as it would be in production, where future data is predicted based on past patterns. Here, we split the dataset into an 80% training set and a 20% test set:

1. Split the data into training and testing sets (80% training, 20% test):

    ```
    train_size = int(len(lagged_data) * 0.8)
    train_data = lagged_data[:train_size]
    test_data = lagged_data[train_size:]
    ```

2. With our features set up in this way, the `Value` column is our target (what we aim to predict), and the other columns (including lagged features) serve as the input features:

```
X_train = train_data.drop('Value', axis=1)
y_train = train_data['Value']
X_test = test_data.drop('Value', axis=1)
y_test = test_data['Value']
```

Now, you have split the dataset with the last 20% of the dataset aside for testing and the earlier 80% to be your training data. In real-world forecasting, however, new data arrives without values for our lagged features, as these are derived from past observations of the target variable. We address this by structuring the model so that, during each prediction step, we can generate lagged features from recent observed values. This method is known as "iterative forecasting" or "rolling predictions" and enables the model to generate new lagged values from each preceding prediction. For now, we'll use the training data to build the model, with these lagged features as predictors. This will enable the model to make predictions based on all available features, without relying on unknown future values in the test set. Later in this chapter, we will discuss how to use this model for future predictions, showing how lagged values are generated in real time as we make forecasts step by step.

In the next section, we will use XGBoost to create a predictive model for the example data.

Training the XGBoost model

With the data prepared, we can now train the XGBoost model. For time series forecasting, XGBoost typically uses the regression objective (`reg:squarederror`), which optimizes for minimizing squared errors. First, we will initialize the XGBoost model and train it on the training data. Then, we can use the model to make predictions on the test set. Let's start:

1. Begin by importing `XGBRegressor` from `xgboost` and initializing the regressor with the objective of `reg:squarederror`. Start with `n_estimators = 100`, `max_depth =3`, and `learning_rate = 0.1`. These are conservative settings compared to the default XGBoost values, which will prevent overfitting of the data. You can tune these hyperparameters later to suit the project you are doing. Details on the XGBoost hyperparameters are in *Chapter 5*:

```
from xgboost import XGBRegressor
model = XGBRegressor(objective='reg:squarederror',
    n_estimators=100, max_depth=3, learning_rate=0.1)
```

2. Next, train the model using the training data: `X_train` and `y_train`:

```
model.fit(X_train, y_train)
```

3. Then, we use the model to predict the target for the test dataset, X_test:

```
y_pred = model.predict(X_test)
```

Here, we used XGBoost to build a model on the training dataset and used that model to predict the target values for inputs in the test dataset. Next, we will learn how to evaluate the results of time series models.

Evaluating the model

Evaluating the model's performance is essential to understanding how well it forecasts future values. For time series models, you can use metrics for regression tasks. Two common metrics are **mean squared error** (**MSE**) and **mean absolute error** (**MAE**). Other metrics are covered in *Chapter 5*. Additionally, you can visualize the actual versus predicted values to see how well the model captures the time series' patterns. Let's try it:

1. Begin by importing the metrics from scikit-learn for mean_squared_error and mean_absolute_error. Prepare for plotting the data by importing matplotlib.pyplot.

2. Next, use the scikit-learn metrics, mean_squared_error and mean_absolute_error, to compare the predicted values, y_pred, to the ground truth values in the test dataset, y_test:

```
mse = mean_squared_error(y_test, y_pred)
mae = mean_absolute_error(y_test, y_pred)

print(f'Mean Squared Error: {mse}')
print(f'Mean Absolute Error: {mae}')
```

We get the following results:

```
Mean Squared Error: 8.721458169486231
Mean Absolute Error: 2.4316627806970614
```

3. Recall from *Chapter 5* that a lower MSE indicates a better fit for the model, and here, an MSE of 8 seems relatively high. To better understand the prediction accuracy, we also calculate the MAE, which measures the average magnitude of errors in the predictions, without considering their direction. The formula for MAE is as follows:

$$MAE = \frac{1}{n}\sum_{i=1}^{n}|y_i - \hat{y}_i|$$

Here, y_i is the actual value, $\hat{y}_i$ is the predicted value, and n is the number of observations. A high MAE suggests substantial differences between predictions and actual values.

4. Lastly, plot the predicted values against the actuals to look for gaps between the prediction and actual values:

```
import matplotlib.pyplot as plt

# Generate the dates for the future predictions (extend the
original date range)
future_dates = pd.date_range(start=X_test.index[-1],
    periods=11, freq='D')[1:]
    # Skip the first date to match steps

# Convert the forecast to a DataFrame for easy plotting
future_df = pd.DataFrame({'Date': future_dates,
    'Predicted_Value': future_forecast})
future_df.set_index('Date', inplace=True)

# Plot the actual and predicted values
plt.figure(figsize=(10, 6))

# Plot actual values from the test set
plt.plot(X_test.index, y_test, label='Actual')

# Plot the future predictions
plt.plot(future_df.index,future_df['Predicted_Value'],
        label='Predicted', linestyle='--')

# Add labels and legend
plt.title('Future Time Series Forecasting with XGBoost')
plt.xlabel('Date')
plt.ylabel('Values')
plt.xticks(rotation=45)
# Rotate X-axis labels by 45 degrees
plt.legend()
plt.show()
```

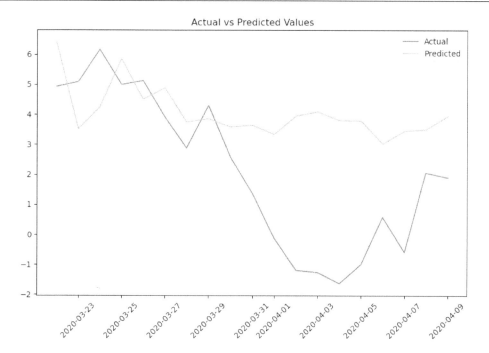

Figure 9.3 – Actual and predicted values for time series data

The plot provides a visual understanding of how well the model is forecasting future values. The closer the predicted values are to the actual values, the better the model's performance. As you can see, for the randomly produced data here, there is a big gap between the actual data and the predicted values in the April 2020 timeframe. In that region, the XGBoost model is not doing a great job of matching the ground truth values. This can happen when something occurs when making predictions that were not seen in the timeframe that was used to train the model. The best way to handle this is to monitor model performance in production and identify triggers for when the model should be retrained.

We've now used XGBoost to build a model based on time series data and make predictions using that model. In the next section, you will learn about predicting multiple future time steps using iterative forecasting.

Forecasting future values

Once we have successfully trained the XGBoost model on historical time series data, the next logical step is to use this model to predict future values. However, in time series forecasting, we often need to predict multiple future time steps. In a supervised learning context, this requires a strategy called **iterative forecasting** or **rolling forecasting**.

What is iterative forecasting?

In iterative forecasting, after predicting the next time step, you use that prediction as an input to forecast the following time step. This process is repeated until you have forecasted all the required future values. This has the benefit of using a single model to generate multiple forecasted values rather than a separate model for each future data point. This method poses two key challenges. The first challenge is a dependency on lagged features. As you generate future predictions, you must update the lagged features with the predicted values. The second challenge is error propagation. Iterative forecasting introduces the risk of error accumulation since each future prediction depends on the accuracy of the previous prediction. If a prediction is inaccurate, the following predictions might also suffer. This difficulty is inherent to the nature of this process. There are advanced techniques, such as deep learning approaches, that can help mitigate this problem.

Forecasting future time steps

To perform the forecasting, we will write a Python function to forecast multiple time steps into the future. This function will take the trained XGBoost model, the most recent data (including lag features), and the number of future steps to forecast. Let's put this into practice:

1. We begin by initializing the input features. Start with the last known data points (lag features from the most recent observations). The `initial_data` argument contains the most recent lagged features. The function extracts the last row (i.e., the latest time step) of this dataset and reshapes it to match the input structure required by the XGBoost model:

    ```
    def forecast_future(model, initial_data, steps=10):
        future_values = []
        input_data = initial_data[-1:].values.reshape(1, -1)
    ```

2. Next, we use the trained XGBoost model to predict the next value based on the current input features. This is done in an iterative forecasting loop. Inside the loop, the function predicts the next time step using `model.predict(input_data)`. We put the predicted values into a `future_values.append(prediction[0])` list, and then repeat for the next prediction:

    ```
    for _ in range(steps):
        # Predict the next value
        prediction = model.predict(input_data)

        # Append the predicted value to the list
        future_values.append(prediction[0])
    ```

3. Then, update the lagged features. Shift the lagged features by one step and insert the newly predicted value as the latest feature for the next iteration. The `np.roll(input_data, -1)` function shifts the input features by one position to the left, simulating the shift in time.

The last element in the input feature array is then replaced with the predicted value to ensure the updated lagged features are used in the next iteration:

```
input_data = np.roll(input_data, -1)
input_data[0, -1] = prediction[0]
```

4. Repeat this process until all required future steps are forecasted. After completing the loop, the function returns a list of all predicted future values:

```
return future_values
```

Now, you've created a function to perform iterative forecasting. In the next section, we will apply this function to forecast 10 future values using the XGBoost model we trained in the earlier sections.

Preparing input data for forecasting

Before using the forecasting function you created in the previous section, you need to prepare the initial input data, which consists of the most recent lagged values from the training dataset. This will serve as the starting point for the forecasting process:

1. First, get the most recent data points for forecasting (lag features). You can do this with X_test.tail(1):

```
latest_data = X_test.tail(1)
```

2. Forecast the next 10 steps by calling the forecast_future function and passing it the model to use, model, the dataset you just prepared, latest_data, and the number of steps, steps=10:

```
future_forecast = forecast_future(model, latest_data,
    steps=10)
```

3. Finish by displaying the predicted future values:

```
print(f"Future Predictions for the next 10 steps: {
    future_forecast}")
```

This results in the following:

```
Future Predictions for the next 10 steps: [3.94813, 2.6983302,
4.077058, 4.6396346, 4.484285, 4.06331, 3.9909317, 4.087828,
4.876842, 4.0725293]
```

Now, you have used iterative forecasting to make future predictions for the next 10 steps. Next, you will graph these predictions.

Visualizing the forecast

To better understand the predictions, let's visualize the future predictions alongside the actual time series data:

1. Prepare to plot the data by importing `matplotlib` and extending the original date range to put dates against future predictions:

    ```python
    import matplotlib.pyplot as plt
    future_dates = pd.date_range(start=X_test.index[-1],
        periods=11, freq='D')[1:]
    # Skip the first date to match steps
    ```

2. Convert the forecast to a DataFrame for easy plotting, and set `index` to be the `Date` column:

    ```python
    future_df = pd.DataFrame({'Date': future_dates,
        'Predicted_Value': future_forecast})
    future_df.set_index('Date', inplace=True)
    ```

3. Plot the actual and predicted values. Plot actual values from the test set, and plot the future predictions using a dashed line:

    ```python
    plt.figure(figsize=(10, 6))
    plt.plot(X_test.index, y_test, label='Actual')
    plt.plot(future_df.index, future_df['Predicted_Value'],
        label='Predicted', linestyle='--')
    ```

4. Add labels and a legend, rotating the x-axis labels by 45 degrees so the dates fit better on the axis:

    ```python
    plt.title('Future Time Series Forecasting with XGBoost')
    plt.xlabel('Date')
    plt.ylabel('Values')
    plt.xticks(rotation=45)
    # Rotate X-axis labels by 45 degrees
    plt.legend()
    plt.show()
    ```

This produces the following graph.

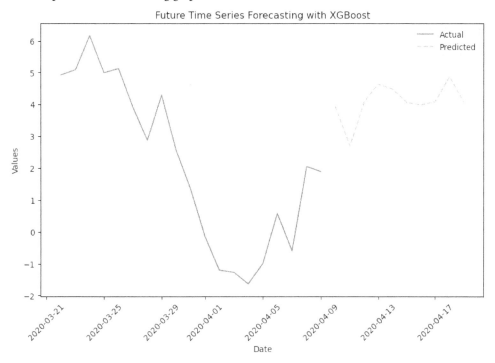

Figure 9.4 – Future time series forecasting

Now, let's interpret these results. The plot shows the actual values from the test set and the predicted future values. The dashed line representing the predicted values should ideally follow the trend set by the actual data. However, it's important to note that since this is an iterative process, errors in the earlier predictions can propagate through future predictions. You will learn about improving future predictions in the next section.

Improving future predictions

While the basic iterative forecasting strategy provides a straightforward way to predict future time steps, there are ways to improve the performance and robustness of the model. First, you can do some feature engineering for the future steps. You can include exogenous variables, such as external factors (e.g., weather, economic indicators), in your time series model to improve the accuracy of future predictions. Your model may also benefit from incorporating more complex seasonal patterns (e.g., daily, monthly) as additional features to help the model capture long-term trends.

Another way to improve future predictions is to use ensemble modeling. Instead of relying on a single model, you can create an ensemble of multiple XGBoost models, each trained with different sets of features or hyperparameters. Averaging the predictions from different models can often yield better performance.

Lastly, you can apply hyperparameter tuning. Fine-tuning the model's hyperparameters (such as the learning rate, number of estimators, and max depth) using techniques such as **grid search** or **random search** can improve the overall accuracy of both short-term and long-term predictions.

Summary

Forecasting future values in time series with XGBoost involves more than just making a single prediction. By iteratively updating the lagged features and using the model to predict multiple future steps, you can extend the forecasting horizon and generate valuable insights for future time periods. However, this process also introduces challenges, such as error propagation, which can be mitigated with better feature engineering and model optimization.

You have now learned about using XGBoost to forecast time series data. You prepared the data by creating lag features and adding date-based features, such as the day of the week, as well as rolling statistics, such as moving averages or rolling means. Then, you learned how to properly split time series data to not lose the inherent ordering. You wrapped up the chapter by predicting values, evaluating the prediction, and forecasting future values. By following these steps and continuously refining the model, XGBoost becomes a powerful tool for time series forecasting, enabling accurate and scalable predictions across a wide range of applications. In the next chapter, you will learn about the importance of model interpretability and explainability with XGBoost, and practice extracting feature importance.

References

- Box, G. E., Jenkins, G. M., & Reinsel, G. C. (2015). *Time Series Analysis: Forecasting and Control* (5th ed.). Hoboken, NJ: Wiley. `https://www.wiley.com/Time+Series+Analysis%3A+Forecasting+and+Control%2C+5th+Edition-p-9781118675021`

- Brockwell, P. J., & Davis, R. A. (2016). *Introduction to Time Series and Forecasting* (3rd ed.). New York, NY: Springer. `https://link.springer.com/book/10.1007/978-3-319-29854-2`

10

Model Interpretability, Explainability, and Feature Importance with XGBoost

While building predictive models, accuracy is often the primary focus. Model interpretability and explainability are good features to have; they are essential for trust, validation, regulatory compliance, and debugging. As machine learning models – particularly ensemble methods such as XGBoost – become more complex, understanding how these models make decisions may not be immediately apparent.

In this chapter, you'll explore the importance of model interpretability and explainability with XGBoost, and practice extracting feature importance.

We'll cover the following topics:

- Why interpretability and explainability matter
- Implementing XGBoost's feature importance
- Exploring SHAP for model interpretation
- Implementing LIME for model interpretation
- Applying ELI5 for model interpretation
- Exploring PDPs for model interpretation

Technical requirements

The code presented in this chapter is available in this book's GitHub repository: `https://github.com/PacktPublishing/XGBoost-for-Regression-Predictive-Modeling-and-Time-Series-Analysis`.

Before diving into the code and explanations, make sure you have all the necessary libraries installed. The recommended method is to install them using `pip`. You'll be using the following Python packages in this chapter:

- shap
- lime
- eli5
- scipy version 1.7.3
- sklearn version 0.24.2

Please note that due to compatibility issues among the lime, SciPy, and scikit-learn libraries, you'll need to install the aforementioned versions of scikit-learn and SciPy.

All examples in this chapter are based on the California housing dataset, which you used previously in *Chapter 4*. This dataset predicts housing price values based on several features.

Why interpretability and explainability matter

Interpretability refers to the degree to which a human can understand the decision that's been made by a model. **Explainability** goes further by providing insights into the specific contribution of each feature to the final prediction. These concepts are vital in many sectors:

- **Healthcare**: Physicians need to understand model predictions to trust them.
- **Finance**: Regulations often require an explanation regarding why a model made a particular decision, especially in lending.
- **Logistics and manufacturing**: Explaining a model's behavior can help improve operational processes, such as inventory management or defect detection, by identifying key drivers.

The XGBoost model itself can become a "black box," meaning that it might be difficult to explain what factors a model used when making a prediction. This is due to XGBoost being an ensemble model where factors from multiple decision trees are combined. This makes it hard to trace a path for a specific prediction. There are tools we can use to shed light on the internal mechanics of XGBoost. In this chapter, you will learn about five such tools. To begin, you'll learn about XGBoost's feature importance scores.

Implementing XGBoost's feature importance

XGBoost automatically provides feature importance scores. These scores indicate how valuable each feature is in constructing the decision trees in the ensemble. XGBoost offers three primary ways to measure feature importance:

- **Gain**: Measures how a feature contributes to the improvement of the model's accuracy, averaged over all trees.

- **Weight**: The number of times a feature appears in the trees (sometimes called *frequency*).

- **Cover**: Measures the relative number of observations related to a feature across all trees.

In this section, you'll use XGBoost's feature importance to understand which factors have the most impact on housing value predictions. You'll do this by loading the dataset, training an XGBoost model, and visualizing the feature importance using XGBoost's built-in methods. You'll use the same California housing dataset you used in *Chapter 4*. Follow these steps:

1. Begin by importing the necessary packages. You'll use XGBoost to model the data. You can use scikit-learn (`sklearn`) to retrieve the California housing dataset and split the data. Finally, you'll use `pandas` to manipulate the data and `matplotlib.pyplot` to make graphs:

   ```
   import xgboost as xgb
   from sklearn import datasets
   from sklearn.model_selection import train_test_split
   from sklearn.preprocessing import StandardScaler
   import pandas as pd
   import matplotlib.pyplot as plt
   ```

2. Next, load the California housing dataset from scikit-learn. You can load it into X and y DataFrames using `return_X_y = True` and `as_frame = True`, respectively:

   ```
   housingX, housingy = datasets.fetch_california_housing(
       return_X_y=True, as_frame=True)
   ```

3. Handle the categorical variables by converting them into numeric form using `pd.get_dummies`. In the case of the `california_housing` dataset, everything is already numeric, so you can skip this step. However, if the dataset you want to work with contains both categorical and numeric data, you'll want to do the following:

   ```
   housing_data_encoded = pd.get_dummies(housingX)
   ```

4. Save the original feature names in a list so that you can use them when you're making a graph of ranked features. This will make the graph easier to interpret.

   ```
   original_feature_names = housing_data_encoded.columns
   ```

5. Split the data into training and testing sets using `train_test_split`. Use the encoded X values, `housing_data_encoded`, and `housingy` for the y values. Setting `random_state = 42` will ensure your results match the ones shown here:

   ```
   X_train, X_test, y_train, y_test = train_test_split(
       housing_data_encoded, housingy, test_size=0.2,
       random_state=42)
   ```

6. Before you train the model, you can do some feature engineering to improve performance. Standardizing the data will help you achieve optimal performance in this case. You can use scikit-learn's `StandardScaler()` to do so:

```
scaler = StandardScaler()
X_train_scaled = scaler.fit_transform(X_train)
X_test_scaled = scaler.transform(X_test)
```

7. Now, you're ready to train the XGBoost model. You can use the squared error loss function for this model by setting `objective = 'reg:squarederror'` as a parameter for `xgb.XGBRegressor`. Then, fit the model:

```
model = xgb.XGBRegressor(objective='reg:squarederror')
model.fit(X_train_scaled, y_train)
```

8. Next, retrieve the feature importance scores and map them back to the original feature names. This allows you to interpret the results of the importance more easily. Remember that feature importance is given as `'gain'` by XGBoost:

```
importance_dict = model.get_booster().get_score(
    importance_type='gain')
```

Then, create a DataFrame for feature importance:

```
pd.DataFrame({'Feature': [original_feature_names[
    int(key[1:])] for key in importance_dict.keys()],
    'Importance': importance_dict.values()})
```

Once you've done this, sort the features in descending order by importance. This makes it easier to interpret the graph of feature importances because they will be displayed with column names and in order from most to least important to the model:

```
importance_dict = model.get_booster().get_score(
    importance_type='gain')
importance_df = pd.DataFrame({
    'Feature': [original_feature_names[int(key[1:])]
    for key in importance_dict.keys()],
    'Importance': importance_dict.values()
})
importance_df = importance_df.sort_values(
    by='Importance', ascending=False)
```

9. Now, you're ready to plot the top 20 most important features, along with their original names:

```
plt.figure(figsize=(10, 8))
plt.barh(importance_df['Feature'].head(20),
    importance_df['Importance'].head(20))
plt.title("Top Important Features (Original Names)")
```

```
plt.xlabel("Importance Score (Gain)")
plt.gca().invert_yaxis()   # Invert y-axis to show the
                           #most important feature on top
plt.show()
```

This will result in the following graph:

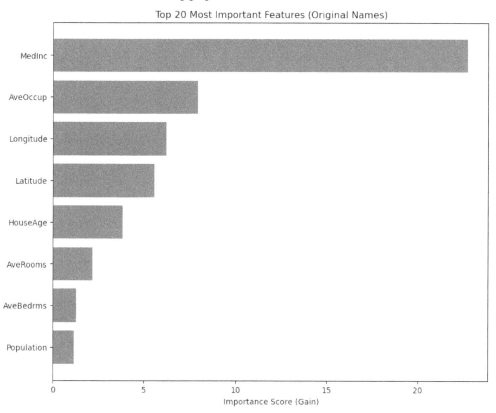

Figure 10.1 – Results from XGBoost performing feature importance

This plot of gain measures the importance score. It shows which features contribute most to the model's predictive performance by calculating the improvement in accuracy for each feature. From this graph, we can see that the most important feature for predicting the median housing value is the median income in the census area, MedInc, followed by the average occupancy (AveOccup), Longitude, and Latitude. This makes sense as we would expect wealthier neighborhoods to have higher housing values; the value of housing is correlated to the number of occupants, which could indicate townhomes or condominium buildings. Here, Latitude and Longitude specify the location of the housing, which is a well-known indicator for real estate value.

The relative ranking of features can help us understand a model's predictions. For instance, AveBedrms (the average number of bedrooms) doesn't seem to be a very important feature in predicting housing value since it's listed second to last of the eight input variables (features). XGBoost's built-in feature importance methods help us get a global understanding of which features are most impactful. However, they don't offer insight into individual predictions or how features interact locally. In the next section, you'll learn about techniques that can provide that insight.

Exploring SHAP for model interpretation

To understand what **SHapley Additive exPlanations** (**SHAP**) is and how it works, imagine that you're playing a team sport, and each player contributes to the team's overall score. Measuring how each team member impacts the score is similar to how SHAP works. SHAP quantifies each feature's contribution to a model's prediction. Based on game theory's **Shapley values**, SHAP provides a consistent and mathematically sound method for explaining both global model behavior and individual predictions.

Why SHAP is useful

SHAP is valuable because it offers consistent, interpretable insights into model behavior. It ensures that if a feature is more influential in one model than in another, it receives a higher contribution score. This approach allows SHAP to give both local and global explanations for feature importance in a model.

Local explanations, also called local importance, help us understand how each feature affects a specific instance's prediction. This means we can interpret why a model made a particular decision for a single case, giving clarity to individual predictions.

Global explanations, known as global importance, provide insights into how each feature influences the model's overall behavior. By showing the average impact of each feature across the dataset, SHAP offers a view of the model's tendencies and dependencies on features in general.

Together, these local and global perspectives offer a full picture of the model's decision-making process, allowing us to interpret both individual predictions and the model as a whole comprehensively and consistently.

Please note that *local importance* and *local explanations*, as well as *global importance* and *global explanations*, are interchangeable terms within SHAP's framework.

Using SHAP with the California housing dataset

Let's use SHAP to understand both global and local model behavior with XGBoost:

1. Begin by importing shap and setting up a DataFrame that uses the scaled X_test values combined with the column names. This allows you to see how the model explains the test dataset:

```
import shap
import pandas as pd
```

```
X_test_scaled_df = pd.DataFrame(X_test_scaled,
    columns=X_test.columns)
```

2. Initialize the SHAP `explainer` function using the `model` object and the scaled training data, `X_train_scaled`. You can set the column names in `X_train` as the feature names:

```
explainer = shap.Explainer(model, X_train_scaled,
    feature_names=X_train.columns)
```

3. Compute SHAP values for the test dataset by calling `explainer` on `X_test_scaled`:

```
shap_values = explainer(X_test_scaled)
```

4. SHAP has a few plotting options. You can graph a SHAP waterfall plot for a single prediction, which doesn't need the `feature_names` parameter. This gives local importance. You'll plot global feature importance in the next step:

```
shap.waterfall_plot(shap_values[0])
```

This results in the following graph:

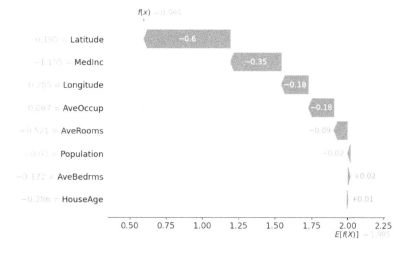

Figure 10.2 – SHAP waterfall plot

This **waterfall plot** explains how the individual features contribute to a *single prediction*. In this case, it explains why a particular house's value, `shap.values[0]`, was predicted by the XGBoost model.

Let's discuss each part of this plot, starting at the top left:

*   The equation, **f(x) = 0.594**, says that the final predicted house price for this particular instance is approximately 0.594. Back in *Chapter 4*, you learned that this dataset uses a value given in units of $100,000. So, the value of this house is predicted to be $59,400 [1990].

- On the far right, you can see the base value – that is, $E[f(x)] = 1.995$ – which is the average predicted house value across the dataset. SHAP breaks down the difference between the base value and the predicted value. The preceding graph shows how each feature contributes either positively (increasing the predicted value) or negatively (decreasing the predicted value).

- The blue bars indicate features that push the prediction lower, while the red bars indicate features that push the prediction higher.

- In this case, the location (`Latitude`, `Longitude`) and the size of the house (`AveOccup`, `AveRooms`) reduce the value, while the number of bedrooms (`AveBedrms`) and the age of the house (`HouseAge`) increase the value.

5. You can use a SHAP summary plot to graph **global feature importance**. This will use the `X_test_scaled_df` DataFrame you created earlier to display all the feature names:

```
shap.summary_plot(shap_values, X_test_scaled_df)
```

This generates a different graph:

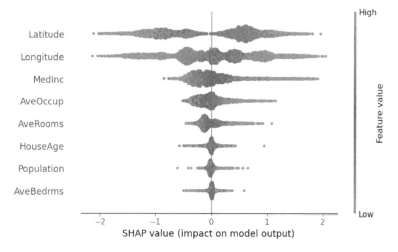

Figure 10.3 – SHAP summary plot

This **summary plot** explains global feature importance across the entire dataset. It provides insights into which features have the most significant impact on model predictions and how different feature values affect the output:

- The features are ranked by their importance (the average absolute SHAP value), from most important at the top to least important at the bottom.

- This graph shows that real estate value really *is* about location, with both `Latitude` and `Longitude` being the most important features in predicting value.

- The X-axis represents the SHAP value that has the most impact on the model's output.

- Values further to the right (positive SHAP values) mean the feature increases the house value. Values further to the left (negative SHAP values) mean the feature decreases the house value.

- Red dots represent high feature values, while blue dots represent low feature values. For example, `AveRooms` has red dots located to the right, showing that larger homes (high `AveRooms`) push the house value higher. Each dot represents one data prediction, and its position on the *X*-axis indicates how much that feature contributed.

- A wide spread (such as for `Latitude` and `Longitude`) means that the feature has a significant impact in different ways for different houses, while a narrow spread (such as `AveBedrms`) suggests that its effect is more consistent across the dataset.

With that, you've used SHAP to look at a model of the California housing dataset. You've seen how SHAP provides both a detailed breakdown of individual predictions and a high-level overview of feature importance. In addition, SHAP values are consistent and reliable, making them highly useful in real-world applications. In the next section, you'll apply LIME and compare the results to those given by SHAP.

Implementing LIME for model interpretation

Local Interpretable Model-agnostic Explanations (**LIME**) uses a different approach from SHAP. While SHAP assigns each feature a contribution score to explain the model's prediction, LIME approaches interpretability by building a simple, local model around the prediction. LIME adjusts the input features in small, controlled ways and observes how these changes influence the prediction, enabling it to construct a straightforward model that explains the prediction in an interpretable manner.

Why use LIME?

LIME is particularly useful for debugging individual predictions or edge cases. It focuses on local explanations. Additionally, LIME works with any machine learning model, not just tree-based models such as XGBoost.

Using LIME to interpret XGBoost predictions

To use LIME effectively, you'll need to pass unscaled data to LIME and use a wrapper function to scale the data before prediction. Let's try it out:

1. To start, import `lime`, `lime_tabular`, `numpy`, and `pandas`. These packages will help you use LIME and present the results:

```
import lime
import lime.lime_tabular
import numpy as np
import pandas as pd
```

2. Next, define a wrapper for `model.predict` to ensure it uses scaled data (`x_scaled = scaler.transform(x)`) and flattens the data (via `return model.predict(x_ scaled).flatten()`) so that you can start using LIME:

```
def predict_fn(x):
    x_scaled = scaler.transform(x)
    return model.predict(x_scaled).flatten
```

3. Now, you can initialize a LIME explainer using the unscaled training data, `X_train.values`. In LIME, continuous variables are often discretized into categories by default, which can cause problems with scaled data. By setting `discretize_continuous=False`, you ensure that LIME uses the actual values. You can use `feature_names=X_train.columns` to have LIME use the column names for each feature, which makes the output easier to understand:

```
explainer = lime.lime_tabular.LimeTabularExplainer(
    X_train.values,
    feature_names=X_train.columns,
    mode='regression',
    discretize_continuous=False
)
```

4. You're now ready to explain a single instance (for example, the first test instance – that is, `i=0`):

```
i = 0
exp = explainer.explain_instance(X_test.values[i],
    predict_fn, num_features=5)
```

5. Finally, show the explanation with `show_in_notebook` and display the table of features and values with `show_table=True`:

```
exp.show_in_notebook(show_table=True)
```

This outputs the following:

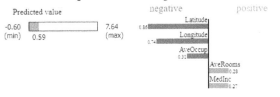

Figure 10.4 – LIME output

LIME's explanation focuses on a specific instance – in this case, the first result (`i=0`). It shows how features such as `Latitude`, `Longitude`, and `AveRooms` affect a single prediction, making it great for debugging unusual predictions. On the left-hand side of *Figure 10.4*, you can see the predicted value for the first result concerning the minimum and maximum predictions – that is, **0.59**. Recall that this is in $100,000 increments, so this housing unit has a value of 59,000. The middle graph shows the effect of five input variables (set using `num_features=5` in *Step 4*). In this case, the location (`Latitude` and `Longitude`) and the average occupancy (`AveOccup`) of this particular instance are on the negative side of the graph. This means they caused the target value to be lower, while the average number of rooms (`AveRooms`) and median income (`MedInc`) were positive influences on the housing value. On the right-hand side of *Figure 10.4*, you can see a table that specifies the feature values that were used for the prediction. Next, we'll compare lea LIME and SHAP.

LIME versus SHAP

SHAP explains the entire model and can explain individual predictions using globally consistent logic. On the other hand, LIME focuses on explaining individual predictions by approximating the model locally. The following table can help you decide which one you should use:

When to use SHAP	You need a global view of how features impact predictions across the entire model.
	You want to explain individual predictions with a consistent feature impact.
	Your use case involves regulatory requirements or demands interpretability that ensures consistent logic for each feature.
When to use LIME	You need to explain a single, specific prediction or decision.
	You don't need globally consistent explanations for all instances.

Table 10.1 – SHAP versus LIME

In this section, you used LIME to understand what factors influence the XGBoost model. This is useful when you wish to quickly understand why a particular decision was made (for example, for product recommendation or loan denial). Next, you'll learn how to apply ELI5.

Applying ELI5 for model interpretation

Explain Like I'm 5 (ELI5)is a tool that was designed to demystify machine learning models and predictions. It provides explanations that are easy to understand, even for complex models such as XGBoost or deep learning models.

Why use ELI5?

ELI5 can show feature importance, explain individual predictions, and help you debug and validate models. ELI5 is particularly useful in industries such as healthcare, finance, and law, where interpretability and transparency are critical. In the next section, you'll use ELI5 to explain housing value predictions.

Explaining an XGBoost model with ELI5

Let's use ELI5 to understand the XGBoost model:

> To start, import `eli5` and `PermutationImportance`. Permutation feature importance is an algorithm that calculates feature importance by measuring how a score (accuracy, R^2, F_1, and so on) is impacted by removing a feature from the model, hence why it's used as the name of the method that calculates feature importance in the `eli5` package:

```
import eli5
from eli5.sklearn import PermutationImportance
```

1. To use the permutation importance algorithm, fit the permutation importance model using `X_test_scaled` and `y_test`. This helps you avoid having to retrain the model while still letting you test what happens to the score when a feature is removed (the algorithm does that part for you):

```
perm = PermutationImportance(model,
     random_state=42).fit(X_test_scaled, y_test)
```

2. Display the feature importance results using `show_weights`. Pass the model fit, `perm`, to the module. Include the feature names via `X_test.columns.tolist()` so that the results are easy to interpret:

```
eli5.show_weights(perm,
     feature_names=X_test.columns.tolist())
```

This outputs the following table of weights and features:

Weight	Feature
1.4170 ± 0.0543	Latitude
1.2683 ± 0.0484	Longitude
0.3664 ± 0.0156	MedInc
0.1504 ± 0.0136	AveOccup
0.1034 ± 0.0074	AveRooms
0.0502 ± 0.0082	HouseAge
0.0083 ± 0.0011	AveBedrms
0.0055 ± 0.0030	Population

Figure 10.5 – ELI5 results

In this table, the `Weight` column gives the importance score associated with each feature, while the `Feature` column displays the name of the input variable that contributes to the model's predictions. A higher weight indicates that the feature has a larger influence on the predictions made by the model. The plus/minus symbol represents the uncertainty or standard deviation around the weight estimate. It gives an indication of the variability in the feature's importance across different instances.

ELI5 provides insight into which features are the most important by presenting a ranked list based on weight. Features with very small weights have minimal contribution to the model, which suggests they're less critical in determining house values in this dataset. For each feature, the uncertainty (the ± value) indicates how consistently the feature contributes to the model's predictions across different instances. A relatively high uncertainty for some features indicates that their importance can vary across individual cases, which is important when interpreting model predictions.

This feature importance analysis helps explain which aspects the XGBoost model focuses on when making predictions and allows us to trust the model's behavior more by understanding the reasoning behind its predictions. Now that you've tried out ELI5, you'll learn how to use PDPs.

Exploring PDPs for model interpretation

When building machine learning models, understanding how individual features impact the model's predictions is helpful, especially for more complex models such as gradient boosting. **Partial dependence plots** (**PDPs** are a valuable tool for visualizing the relationship between a feature and the target variable, holding all other features constant. PDPs allow you to explore how varying a single feature influences the predictions made by the model.

Why use PDPs?

PDPs help in visualizing the marginal effect of one or two features on the predicted outcome. They're particularly useful for understanding how specific features affect predictions. They also help with observing whether a relationship between the feature and the target is linear, monotonic, or more complex. Like SHAP, PDPs provide a global view of the feature's influence on the model across all instances, unlike local interpretability methods such as LIME, which focus on a single instance.

Explaining gradient boosting model features with PDPs

In this example, you'll use scikit-learn's `GradientBoostingRegressor` to train a model and then use PDPs to interpret the model's behavior for the top three most important features. Follow these steps:

1. Start by importing `GradientBoostingRegressor`, `datasets`, `train_test_split`, and `plot_partial_dependence` from scikit-learn. You'll also need `matplotlib.pyplot` and `pandas` to make graphs of the results:

    ```
    from sklearn.ensemble import GradientBoostingRegressor
    from sklearn import datasets
    from sklearn.model_selection import train_test_split
    from sklearn.inspection import plot_partial_dependence
    import matplotlib.pyplot as plt
    import pandas as pd
    ```

2. Load the housing dataset and split the data into training and testing sets:

    ```
    housingX, housingy = datasets.fetch_california_housing
    (return_X_y=True, as_frame=True)
    X_train, X_test, y_train, y_test = train_test_split(
    ```

```
        housingX, housingy, test_size=0.2,
        random_state=42)
```

3. Train the gradient boosting model using `GradientBoostingRegressor`. Then, fit the model on the training dataset:

```
gbr = GradientBoostingRegressor(n_estimators=100,
    learning_rate=0.1, max_depth=5, random_state=42)
gbr.fit(X_train, y_train)
```

4. Get `feature_importances` values from the gradient boosting model and put them into a list that you can convert into a DataFrame. This will let you pair them with the feature names for easier interpretation:

```
importances = gbr.feature_importances_
feature_importances_df = pd.DataFrame({
    'Feature': X_train.columns,
    'Importance': importances
})
```

5. Sort the features by importance in descending order and select the top six using `head(6)`. We're using an even number here so that the plots can be in two rows for readability:

```
top_features = feature_importances_df.sort_values(
    by='Importance', ascending=False).head(5)
print("Top 6 Important Features:")
print(top_features)
```

This results in the following output:

```
Top 6 Important Features:
      Feature  Importance
0      MedInc    0.578121
5     AveOccup    0.132026
7    Longitude    0.105324
6     Latitude    0.096638
1     HouseAge    0.043179
2     AveRooms    0.025655
```

Now, you can make graphs of the partial dependence for each of the top six features.

6. To start making these graphs, get the indices of the top six features by looping over the list of `top_features` and using `get_loc`:

```
top_features_indices = [X_train.columns.get_loc(
    feature) for feature in top_features['Feature']]
```

7. Then, generate PDP plots for the top six features using `plot_partial_dependence`. This method takes the model, the training dataset, the indices of the top features, and the list of feature names. Show the plots with `plt.show()`:

```
fig, ax = plt.subplots(nrows=3, ncols=2,
    figsize=(12, 12), sharey = True)
plot_partial_dependence(gbr, X_train,
    top_features_indices,
    feature_names=X_train.columns, grid_resolution=50,
    ax=ax)
plt.show()
```

Here are the plots:

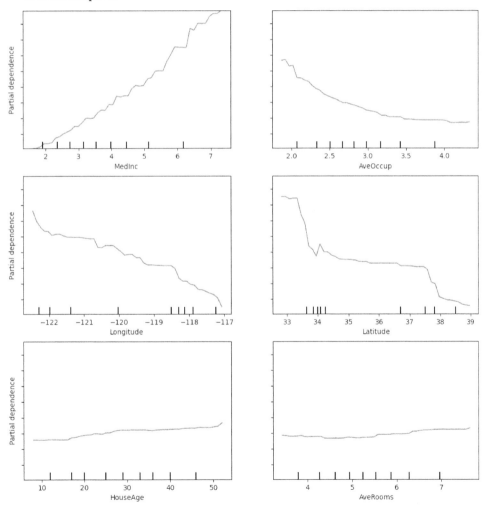

Figure 10.6 – PDPs

These plots show the relationship between a feature and the average predicted outcome of the model, with the feature having the values shown on the X-axis if you hold all the other features constant. A positive relationship indicates that an increase in the feature value increases the predicted output, such as in the case of median income (`MedInc`), where the top-left graph shows a rising partial dependence with an increased `MedInc` value. PDPs can show non-linear relationships between the input variables and the output. For example, regarding average occupancy, there's a non-linear (curved) relationship between `AveOccup` and the prediction. The bars along the X-axis indicate the distribution of the data in that input column.

> **Why gradient boosting and not XGBoost?**
>
> Due to some compatibility issues between the XGBoost API and scikit-learn's `plot_partial_dependence` function, we used scikit-learn's own gradient boosting implementation, which integrates seamlessly with the `plot_partial_dependence` function. This approach ensured that we generated accurate and interpretable PDPs without encountering API issues.

Summary

Understanding and explaining complex machine learning models such as XGBoost is necessary for ensuring transparency, trust, and effective debugging. In this chapter, you explored a range of interpretability techniques – SHAP, LIME, ELI5, XGBoost's feature importance, and PDPs – each providing valuable insights into how the model works.

First, you saw how SHAP excels at both global and local explanations, providing a comprehensive view of how features influence the model overall, as well as detailed, instance-specific insights. Then, we looked at LIME, which offers flexible, model-agnostic explanations while focusing on individual predictions and revealing what drives each model output. Next, we covered ELI5, which simplifies global feature importance, offering intuitive explanations that make it easier to communicate key model drivers.

In addition to these tools, you used XGBoost's built-in feature importance, which offers a clear ranking of the most influential features, helping you prioritize the factors that most impact predictions. The final tool you explored, PDPs, add further clarity by visualizing the relationships between features and predictions, particularly in capturing nonlinear effects.

Incorporating these tools elevates your ability to not only create high-performing models but also to explain and trust their decisions. By making the inner workings of these complex models accessible and transparent, you ensure that your work has both impact and accountability. By using SHAP, LIME, ELI5, XGBoost's feature importance, and PDPs, you can bridge the gap between cutting-edge technology and real-world understanding, creating models that can be trusted and applied with confidence.

In the next chapter, you'll learn about metrics that you can use to evaluate models and compare model performance.

Part 3:
Model Evaluation Metrics and Putting Your Model into Production

In this part, you will expand on the model metrics used in *Part 2* and measure how well a model is working for your dataset and how to adjust modeling parameters to improve the measurements, including metrics for classification models. You will learn how to use pipelines to manage feature engineering tasks. Lastly, you will gain experience in deploying an XGBoost model into a production environment.

This part contains the following chapters:

- *Chapter 11, Metrics for Model Evaluations and Comparisons*
- *Chapter 12, Managing a Feature Engineering Pipeline in Training and Inference*
- *Chapter 13, Deploying Your XGBoost Model*

Metrics for Model Evaluations and Comparisons

In this chapter, we'll measure how well a model is working for your dataset and how to adjust modeling parameters to improve the predictions. First, we will get hands-on experience with scikit-learn and the Python APIs for XGBoost. Then, we will learn about hyperparameter tuning and using metrics to measure how well a model is working. We learned about using some of these model-fitting metrics (such as R^2 and RSME) in *Chapter 5*. We will expand on that understanding to include metrics for evaluating classification models in this chapter. Lastly, we will share cautions on over- and underfitting.

In this chapter, we will cover the following main topics:

- Working with the XGBoost API

- Evaluating model performance for classification and regression

- Tuning XGBoost hyperparameters to improve model fit and efficiency

Technical requirements

You can use the same virtual environment which is already set up with the necessary packages from earlier chapters of this book. The code presented in this chapter is available on our GitHub repository:

`https://github.com/PacktPublishing/XGBoost-for-Regression-Predictive-Modeling-and-Time-Series-Analysis`

You will use the following software:

- Python 3.9

- XGBoost 1.7.3

- NumPy 1.21.5

- pandas 1.4.2

- scikit-learn 1.4.2

- Seaborn

- Anaconda

- VS Code

Working with the XGBoost API

There are two ways to use XGBoost with Python: the native API and the scikit-learn API. The primary difference between these two methods is that the native API requires you to convert your data into a DMatrix. A DMatrix is a specialized data structure used by XGBoost to optimize both memory usage and computation speed during model training. It stores data in a format that allows XGBoost to efficiently perform tasks such as sparse matrix optimization, enabling faster training on large datasets. The DMatrix format is particularly useful for handling missing values, as it can store sparse data without wasting memory on missing entries. You'll need to convert your datasets into this format when using XGBoost's native API for greater control over training parameters and performance.

The scikit-learn API works with pandas DataFrames and NumPy arrays. So far, in previous chapters, you've used the scikit-learn API. However, it abstracts some things (in other words, it's doing some stuff for you) that you might wish to have more control over. That's where the native API can offer additional flexibility.

Using the native API

With XGBoost's native Python API, you must first convert your data into DMatrix format, which is a more efficient data structure for large datasets. Using the native API allows you full access to all XGBoost's features without relying on updates from external libraries like scikit-learn. To train the model, you call `xgb.train`, while in the scikit-learn API, you would use `modelname.fit`. Similarly, to make predictions with the native API, your input data needs to be in DMatrix format.

Using the scikit-learn API

In previous chapters, you used the scikit-learn API to call XGBoost. This API allows you to work with XGBoost in a familiar format, similar to other models in scikit-learn. For example, to train a model, you would use `modelname.fit`, and to make a prediction, you would use `modelname.predict`. The scikit-learn API makes it easier to switch between different models while evaluating which one works best for your dataset. However, if you decide to fully commit to XGBoost, switching to the native API might offer more control and additional customization options not available through the scikit-learn wrapper.

Comparing the scikit-learn and native Python APIs

To sum up, the scikit-learn API is a bit easier to use and allows integration into scikit-learn pipelines. You should use the XGBoost native API if you need maximum performance because it provides more flexibility and full control over the model you are building. To see how they compare, you can build a model on the same dataset with first the scikit-learn API and then with the XGBoost native API. Let's do that now:

1. To start, you will need to import the following packages: xgboost, pandas, and train_test_split from scikit-learn. You can use the California housing dataset, which is also from scikit-learn, for these models:

```
import xgboost as xgb
import pandas as pd
from sklearn.model_selection import train_test_split
from sklearn.datasets import fetch_california_housing
```

2. Next, we load the dataset and split the data. This is the same code we wrote in *Chapter 4*:

```
housingX, housingy = fetch_california_housing (
    return_X_y=True, as_frame=True)
X_train, X_test, y_train, y_test = train_test_split(
    housingX,housingy, test_size=0.2, random_state=17)
```

3. Now, you can use the scikit-learn API for XGBoost to train a model, xgb.XGBRegressor, with the training dataset, model.fit(X_train, y_train), and make a prediction for the test dataset, model.predict(X_test), just as you've done in *Chapter 4*. To be certain you are making a straight comparison, set the following model parameters: max_depth = 3, learning_rate=0.1, and n_estimators=100:

```
sklearn_model = xgb.XGBRegressor(
    objective='reg:squarederror', max_depth=3,
    learning_rate=0.1, n_estimators=100)
sklearn_model.fit(X_train, y_train)
sklearn_preds = sklearn_model.predict(X_test)
```

4. To use the native Python API for XGBoost, you will first need to put the data into the DMatrix format:

```
dtrain = xgb.DMatrix(X_train, label=y_train)
dtest = xgb.DMatrix(X_test, label=y_test)
```

5. Now you can set up the parameters for the model. Be sure to use the same parameters you used for the scikit-learn model to allow for a fair comparison:

```
params = {
    'objective': 'reg:squarederror',
    'max_depth': 3,
    'learning_rate': 0.1
}
```

6. To train the model with the native Python API, we use xgb.train and pass the parameters, the training set, and the number of rounds, num_boost_round, which is the Python API analog to n_estimators in the scikit-learn API:

```
xgbapi_model =xgb.train(params, dtrain,
    num_boost_round=100)
```

7. You can now make predictions with the model with xgbapi_model.predict. Remember that you need the input data for the prediction to be in DMatrix format, but you took care of that earlier:

```
xgbapi_preds = xgbapi_model.predict(dtest)
```

8. To compare the output, you can make an x-y plot with the native Python API predictions, xgbapi_preds, on the y-axis and the scikit-learn API predictions, sklearn_preds, on the x-axis. If they match, you will see a straight diagonal line. You can use matplotlib. pyplot to make this plot with plt.scatter. To make the points more visible, you can play around with the transparency using alpha = 0.5, and make the edge or border of each point white with edgecolors = 'white'. If you'd like, you can add a title and axis labels:

```
import matplotlib.pyplot as plt
plt.figure(figsize=(10, 6))
plt.scatter(sklearn_preds, xgbapi_preds, alpha=0.5,
    edgecolors='white')
plt.xlabel('Scikit-learn Predictions')
plt.ylabel('XGBoost API Predictions')
plt.title('Scatter Plot of Predictions:
    Scikit-learn vs XGBoost API')
plt.show()
```

This produces the following plot:

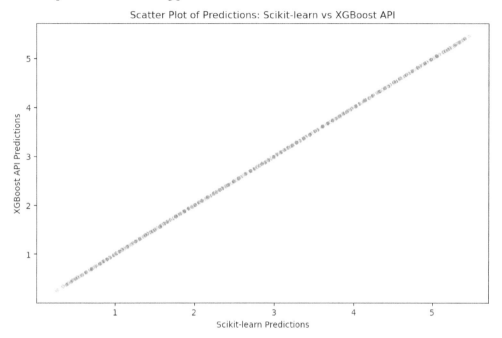

Figure 11.1 – Scatter plot of predictions from different XGBoost APIs,
comparing the XGBoost API to the scikit-learn API

This graph shows that XGBoost models created with the scikit-learn API and the native Python API produce the same predictions, as evidenced by the fact that the points, plotted with scikit-learn API predictions on the x-axis and the native Python API predictions on the y-axis, fall on a diagonal.

To sum up, the scikit-learn API is easier to use and integrates well into scikit-learn pipelines, making it ideal for users who are already familiar with scikit-learn and want to quickly train, evaluate, and switch between different machine learning models. It abstracts much of the complexity behind a high-level interface, making it suitable for experimentation and prototyping. For instance, when you need to quickly compare XGBoost to other models such as Random Forest or Linear Regression, the scikit-learn API provides a consistent, familiar interface.

On the other hand, the XGBoost native API offers more control and flexibility. This is important when you need to fully exploit XGBoost's advanced features and customize the training process beyond what the scikit-learn API allows. For example, the native API supports distributed training, finely tuned parameter control, and better memory optimization using the DMatrix format. The native API is also frequently updated with new features, so if you're working on cutting-edge projects or need to optimize performance for large-scale datasets, the native API might be the better choice.

In general, here are some guidelines:

- If you need ease of use and compatibility with other models, stick to the scikit-learn API
- If you need advanced features and maximum performance with fine-grained control over your model, switch to the native API

For your projects, consider starting with the scikit-learn API for simplicity and ease of comparison with other models. Then, if you need more control over the model training process or want to deploy the model in a high-performance environment, you can transition to the native API.

In this section, you practiced with the native XGBoost API for Python and learned about the differences between it and the scikit-learn API, which you used in previous chapters. Next, you will learn about evaluating the performance of models.

Evaluating model performance of classification and regression models

In *Chapter 5*, you compared various models and assessed their performance using **Root Mean Square Error** (**RMSE**), R^2, the time needed to train the model, and the time needed to calculate a prediction. That chapter focused on regression models. In this section, you will learn about the metrics to use for comparing models when performing classification tasks. To make a complete view and enable the use of this chapter as a reference in the future, we'll include the metrics for regression models as well.

The reason different metrics are needed for the different tasks is that there are different types of data. In the case of a classification task, the model predicts which group a particular item belongs to. A regression model predicts the value of a target parameter. To measure how well the model predicts what is happening for these different tasks, different measurements are required. In this section, you will start with a simple classification task called **binary classification**, which determines whether something is in a group or not. Then you will progress to classification for multiple groups. The section will end with a review of the metrics used to assess regression models.

Binary classification

In binary classification, there are only two choices for the model to make. The task of the model is to predict, for an item with a given set of parameters, whether an item belongs to the group or not. A common type of binary classification is defect detection. The question this type of model answers is: does this part have a defect? The two choices are yes, there is a defect (or **true**), and no, there is no defect (or **false**).

However, what if the model is wrong? Since there are two classifications, the model can be wrong in two ways. It can mark something that doesn't have a defect as having a defect, or it can mark something with a defect as not having a defect. The first case is a *false positive* or *Type I error*. The second case is a *false negative* or *Type II error.* This table (*Table 11.1*) summarizes the types of error for a binary classification model:

Actual (ground truth)	Prediction (model output)	
	No defect	Defect
No defect	True Negative; Correct	False Positive; Type I error
Defect	False Negative; Type II error	True Positive; Correct

Table 11.1 – Table of binary classifier error types

To assess a model's accuracy, we can compare the rates of both types of error and create metrics that measure how often a model will falsely flag good material (*Type I error*) or pass along bad material (*Type II error*). Depending on the task, you may prefer a model that is more likely to be wrong in one direction rather than the other.

For example, in a medical application such as cancer detection, it's often more important to detect potential cases, even if it means generating false positives (Type I errors). The cost of a false negative—failing to detect cancer when it is present—could be much higher, as it may delay crucial treatment. Therefore, you might design the model to err on the side of caution, accepting more false positives to reduce the risk of missing a real case (Type II error).

This tradeoff between error types is important, as some tasks prioritize avoiding one type of mistake over the other. For this reason, the metrics used to assess binary classifiers give you insights into both types of errors as well as the model's overall accuracy, allowing you to tailor the model to fit the needs of the specific task. Next, you'll learn about the metrics to assess a binary classifier.

Binary classifier metrics – accuracy, precision, recall, and F_1 score

There are four commonly used metrics for assessing a binary classifier: accuracy, precision, recall, and F_1 score. The reason for there being four metrics is to let you select models depending on the problem's requirements. Remember the example from earlier; depending on the problem, you may prefer that if the model is wrong, it be wrong in a particular way. For example, you may want the cancer classifier to warn there may be cancer when there isn't because that is a "safer" choice for the model to make. To work with these metrics, let's say we have a defect classifier. It has predicted the classification of 20 items, 10 with defects and 10 without defects, as having a defect or not according to the following table (*Table 11.2*):

Defect classifier	Prediction	
Actual	No defect	Defect
No defect	7 True Negatives	3 False Positives
Defect	1 False Negative	9 True Positives

Table11.2 – Defect classifier example

Let's start with **accuracy**, since it is the most intuitive of the four metrics. Accuracy is defined as the number of correct predictions divided by the total number of predictions. This is given as the sum of the number of true positives and true negatives divided by the sum of the number of **true positives** (**TP**), **true negatives**(**TNs**), **false positives** (**FP**), and **false negatives** (**FN**)s. That bottom number is just the size of the dataset, or N.

$$accuracy = \frac{(TP + TN)}{(TP + TN + FP + FN)} = \frac{(TP + TN)}{N}$$

For our defect classifier, the *accuracy* is given as *(9+7)/ (9+7+3+1) = 16/20 = .8 =80%*.

To calculate accuracy in Python, you can use the built-in `accuracy_score` function within the scikit-learn metrics model. You can build these examples in a Jupyter notebook to see how they work on our defect classifier example:

1. Prepare the defect classifier data. Set `y_pred` and `y_true` as shown here:

   ```
   y_pred = [0,0,0,0,0,0,0,1,1,1,0,1,1,1,1,1,1,1,1,1,1]
   y_true = [0,0,0,0,0,0,0,0,0,0,1,1,1,1,1,1,1,1,1,1,1]
   ```

2. Import the `metrics` module from scikit-learn and calculate the accuracy using the `accuracy_score` function:

   ```
   from sklearn.metrics import accuracy_score
   accuracy_score(y_true, y_pred)
   print ("Accuracy = ", accuracy)
   ```

   This results in the following output:

   ```
   Accuracy = 0.8
   ```

Now you've seen how to calculate accuracy, let's look at **precision** and **recall**.

Precision is defined as how many of the positive predictions made by the model were correct. This is calculated as the number of TPs divided by the sum of the number of TPs and FPs.

$$precision = \frac{TP}{(TP + FP)}$$

For our defect classifier example, the *precision* is *9 / (9+3) = 9/12 = 0.75 = 75%*.

Recall is defined as how many correct positive predictions were made by the model as a fraction of all the positive cases. This is calculated as the number of TPs divided by the sum of the number of TPs and FNs.

$$recall = \frac{TP}{(TP + FN)}$$

For the defect classifier, the *recall* is *9 / (9+1) = 9/10 = 0.9 = 90%*.

Next, you will use Python and scikit-learn to calculate precision and recall for the defect classifier example using `precision_score` and `recall_score` from scikit-learn:

```
from sklearn.metrics import precision_score, recall_score
precision = precision_score(y_true, y_pred)
recall = recall_score(y_true, y_pred)

print ("Precision = ", precision)
print ("Recall = ", recall)
```

This results in the following:

```
Precision = 0.75
Recall = 0.9
```

Precision and recall are used together because they aren't particularly useful separately. You can have perfect precision so long as everything marked as a defect is actually a defect. However, this doesn't say anything about defects the model missed. Similarly, perfect recall means that every defect was actually a defect but doesn't say anything about false positives – things that are not defects that were tagged by the model as defects. These reasons are why precision and recall are used together. Further, the **F₁ score** combines precision and recall into a single metric. F_1 score is calculated as follows:

$$F_1 = 2 \cdot \frac{precision \cdot recall}{precision + recall}$$

The F_1 score places equal weight on precision and recall. The more general form of this score is called F-beta and allows adjustment of the weighting between precision and recall. The combination of precision and recall makes the F_1 score useful when the dataset is imbalanced and there are fewer examples of one case than the other. If your dataset has only a handful of samples with defects, your model might be trained to always say no defect. Using both accuracy and F_1 score will help detect that case. If a classifier says everything is no defect when there are a small number of defects, the accuracy might be high, but the F_1 score will come back as nan.

The F_1 score for the defect classifier example is as follows:

F1 = 2 ((0.75) (0.9)) / (0.75+0.9)

= 2 (0.675 / 1.65)

= 2 (0.409)

= 0.818 = 81.8%

Next, you'll write Python code to calculate the F_1 score for the defect classifier example. Scikit-learn has an F_1 score in the `metrics` module. You can use it to make the calculation. Since the value is repeating, We've chosen to format the result with three values after the decimal:

```
from sklearn.metrics import f1_score
f1 = f1_score(y_true, y_pred)
print(f"F1 Score = {f1:.3f}")
```

This prints out the following:

```
F1 Score = 0.818
```

While accuracy is more straightforward, the F_1 score provides more insight into how well your model is doing if you have an imbalanced dataset.

Evaluating a defect classifier on imbalanced data

When dealing with imbalanced datasets—where one class significantly outweighs the other—accuracy alone is often insufficient to measure the effectiveness of a machine learning model. This is particularly true in cases such as defect detection, where the number of defective products (positive class) is far smaller than the number of non-defective products (negative class).

To better understand how imbalanced data can impact model evaluation, we will calculate and compare accuracy, precision, recall, and F_1 score for two defect classification scenarios. In the first scenario, we have few defects and many non-defective examples, while in the second scenario, the data is flipped and defects are more common than non-defective examples.

Scenario 1 – few defects and many non-defective examples

In this case, the model is trained on an imbalanced dataset where defects are rare, and most examples are non-defective. The confusion matrix for the defect classifier is shown here (*Table 11.3*).

Defect classifier	Prediction (no defect)	
Actual	No defect	Defect
No defect	18 TNs	0 FPs
Defect	1 FN	1 TP

Table 11.3 – Confusion matrix for a defect classifier with imbalanced
data, where most examples are no defect

Let's calculate the accuracy, precision, recall, and F_1 score for a model dealing with imbalanced data (where defects are rare). First, define the true and predicted values for the dataset, then use scikit-learn's metrics to compute these performance measures. Finally, print out the results:

```
y_true_imbalanced = [0, 0, 1, 1, 0, 0, 0, 0, 0, 0, 0, 0, 0,
    0, 0, 0, 0, 0, 0, 0]
y_pred_imbalanced = [0, 0, 1, 0, 0, 0, 0, 0, 0, 0, 0, 0, 0,
    0, 0, 0, 0, 0, 0, 0]

# Calculate performance metrics
accuracy_imbalanced = accuracy_score(y_true_imbalanced,
    y_pred_imbalanced)
precision_imbalanced = precision_score(y_true_imbalanced,
    y_pred_imbalanced)
recall_imbalanced = recall_score(y_true_imbalanced,
    y_pred_imbalanced)
f1_imbalanced = f1_score(y_true_imbalanced,
    y_pred_imbalanced)
print("Accuracy, precision, recall, F1 score for imbalanced
classifier")
print(f"Accuracy = {accuracy_imbalanced:.3f}")
print(f"Precision = {precision_imbalanced:.3f}")
print(f"Recall = {recall_imbalanced:.3f}")
print(f"F1 Score = {f1_imbalanced:.3f}")
```

This results in the following output:

```
Accuracy, precision, recall, F1 score for an imbalanced classifier
Accuracy = 0.950
Precision = 1.000
Recall = 0.500
F1 Score = 0.667
```

In this case, while the accuracy is high at 95%, it is somewhat misleading because the model only detected one out of two defects. The precision of 1.0 means that when the model predicted a defect, it was always correct. However, the recall (0.5) shows that the model missed 50% of the actual defects. The F_1 score, which balances precision and recall, is 0.667, indicating there is room for improvement in the model's ability to capture all defects.

Scenario 2 – many defects and few non-defective examples

Now, let's flip the data imbalance and assume most examples are defects, with only a few non-defective cases This scenario is illustrated in *Table 11.4*:

Defect classifier	Prediction (mostly defects)	
Actual	No defect	Defect
No defect	1 TN	1 FP
Defect	0 FNs	18 TPs

Table 11.4 – Confusion matrix for a defect classifier with imbalanced
data, where most examples are defects

Let's calculate the accuracy, precision, recall, and F_1 score for a model dealing with imbalanced data (mostly defects). First, define the true and predicted values for the dataset, then use scikit-learn's metrics to compute these performance measures. Finally, print out the results:

```
y_true_defects = [1, 1, 0, 0, 1, 1, 1, 1, 1, 1, 1, 1, 1, 1,
    1, 1, 1, 1, 1, 1]
y_pred_defects = [1, 1, 0, 1, 1, 1, 1, 1, 1, 1, 1, 1, 1, 1,
    1, 1, 1, 1, 1, 1]

# Calculate performance metrics
accuracy_defects = accuracy_score(y_true_defects,
    y_pred_defects)
precision_defects = precision_score(y_true_defects,
    y_pred_defects)
recall_defects = recall_score(y_true_defects,
    y_pred_defects)
f1_defects = f1_score(y_true_defects, y_pred_defects)
print("Accuracy, precision, recall, F1 score for defect classifier")
print(f"Accuracy = {accuracy_defects:.3f}")
print(f"Precision = {precision_defects:.3f}")
print(f"Recall = {recall_defects:.3f}")
print(f"F1 Score = {f1_defects:.3f}")
```

This gives the following output:

```
Accuracy, precision, recall, F1 score for an imbalanced classifier
Accuracy = 0.950
Precision = 0.947
Recall = 1.000
F1 Score = 0.973
```

Here, the accuracy remains high at 95%, and both the precision (0.947) and recall (1.000) are also high. The F_1 score is 0.973, indicating that the model is performing well in this scenario. However, the model did miss one of the two non-defective cases, which is important to note. Even though the accuracy, precision, and recall are all high, the model's focus on identifying defects could lead to missing a small number of non-defective cases.

Key takeaways

When evaluating a model trained on imbalanced data, it's important to use precision, recall, and F_1 score, in addition to accuracy. While accuracy may appear high, precision and recall provide more insight into how well the model is identifying positive cases (defects). In cases where defects are rare, high precision but low recall could indicate the model is overly conservative in predicting defects. On the other hand, when defects are more frequent, high recall with slightly lower precision might indicate the model is prone to false positives.

To handle imbalanced data effectively, it's crucial to align the evaluation with the business problem. In most cases, you'll want to ensure that the positive class (defects) is well-represented in the metrics. If missing defects is costly, maximizing recall and F_1 score can be more beneficial than simply optimizing for accuracy.

In the next section, we'll explore graphical assessments of binary classifiers to further analyze performance.

Graphical analysis of binary classification, ROC-AUC curves, and confusion matrix

Using graphs can often help our understanding, so next, we'll make some plots that visualize how well the model is predicting the actual values. The first graph you will make is called the **Receiver Operating Characteristics Curve (ROC)**. It has an associated metric, the **Area Under the Curve (ROC-AUC)**, or sometimes just **AUC**. This measures how well a model can differentiate between classes. The higher the AUC, the better the model is at distinguishing between the classifications. Let's plot the graph.

Plot the ROC curve and calculate the AUC using the scikit-learn metrics module. In fact, it will draw a very nice graph and automatically calculate the AUC:

```
from sklearn.metrics import RocCurveDisplay
RocCurveDisplay.from_predictions(y_true, y_pred)
```

This makes the following plot:

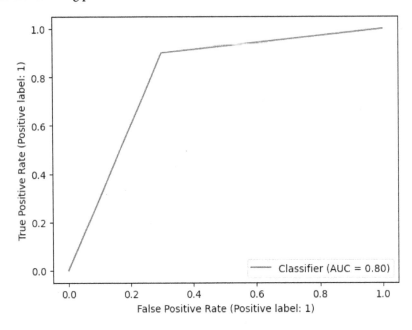

Figure 11.2 – Receiver operating characteristic curve for the defect classifier example

Better models produce a curve closer to the top left corner of the ROC graph, with a higher AUC. A model that randomly associates the samples into classes, a "pure guess," makes a diagonal line through the middle of the graph from bottom left to top right. This defect classifier is doing a good job, with an AUC of 0.8.

The other visual to use is a confusion matrix. This shows how many samples of the data belong to each class compared with where the model classified them:

```
# Generate the confusion matrix
cm = confusion_matrix(y_true_defects, y_pred_defects)

# Create the confusion matrix display
disp = ConfusionMatrixDisplay(confusion_matrix=cm,
    display_labels=["No Defect", "Defect"])
# Plot the confusion matrix using the 'Greens' colormap
disp.plot(cmap=plt.cm.Greens)
plt.title("Confusion Matrix for \nDefect Classification")
plt.xlabel("Predicted Class")
plt.ylabel("True Class")
```

```
# Add custom text legend to show the 0 → No Defect and 1 → Defect
mapping
plt.text(3, 0.1, "Legend:", fontsize=10, weight='bold',
    color='black')
plt.text(3, 0.3, "0 → No Defect", fontsize=9,
    color='black')
plt.text(3, 0.5, "1 → Defect", fontsize=9, color='black')

# Show the plot
plt.tight_layout()
plt.show()
```

This makes the following graph:

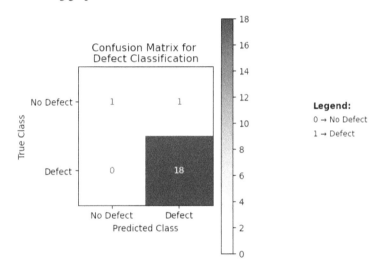

Figure 11.3 – Confusion matrix for the defect classifier example

The visual shows a comparison of predicted classifications to the actual ground truth classifications. The number of TNs, FPs, FNs, and TPs are provided in the matrix. Ideally, the top-right (FPs) and bottom-left (FNs) boxes would contain 0, indicating perfect classification.

The confusion matrix is particularly useful when you want to understand the breakdown of prediction outcomes across both positive and negative classes. This is especially important when dealing with imbalanced datasets, as it allows you to see whether the model is biased toward predicting one class over the other.

In general, you should use a confusion matrix when you need more insight than a single metric (such as accuracy) can provide. It helps you visualize the types of errors the model is making and can be a better choice for evaluating classification models in cases where class imbalance or specific error types (e.g., FPs) are important.

Next, you'll learn how to calculate classifier metrics when there are more than two classifications.

Multi-class classifier metrics – micro, weighted, and macro averages

If you have more than a binary classification problem, say (for example) you are grading tomatoes and they can be poor, fair, good, or excellent, then there is an additional layer to be added. This layer is how to divide up the problem into a series of binary classification problems. There are two approaches: **One-Versus-One (OVO)** or **One-Versus-Rest (OVR)**. *Table 11.5* shows the groupings for the tomato classification example for both OVO and OVR.

Tomato grading	Classifiers
OVO	Poor versus Fair
	Poor versus Good
	Poor versus Excellent
	Fair versus Good
	Fair versus Excellent
	Good versus Excellent
OVR	Poor versus [Fair, Good, Excellent]
	Fair versus [Poor, Good, Excellent]
	Good versus [Poor, Fair, Excellent]
	Excellent versus [Poor, Fair, Good]

Table 11.5 – Table of classifiers for the tomato grading example

As seen in *Table 11.5*, there are only four comparisons that need to be made for the OVR case, compared to six for the OVO case. In general, OVR requires less computational power, so it is more commonly used. It is not perfect, however, because it makes each binary problem imbalanced. By making only one class at a time positive and all the rest of the data negative, you always have a mismatch in the sizes of the classification groups. You can address this by looking at more than just accuracy to evaluate your models, as discussed previously.

Since in the OVR case, we are breaking the problem down into multiple classifiers, we need a way to combine the metrics into a single value for the full model. The **micro average** is the same as accuracy. The **weighted average** takes the class size into account when calculating the average. The **macro average** is the mean of the scores for all of the classifiers. As you saw in the example with mostly defects in the *Binary classifier metrics – accuracy, precision, recall, and F_1 score* section, accuracy can be misleading if classes are imbalanced. For this reason, the micro average is rarely used.

To see how this works, you can use the tomato classification example to explore the metrics for multiclass classifiers:

1. Plot the confusion matrix for our tomato classifier example. You can use the
    ConfusionMatrixDisplay function from scikit-learn. Start by setting the true and
    predicted values for the tomato example. We're pretending a classifier has predicted these
    results from data:

```python
# Define the grades (class labels) and corresponding true/
predicted values
grades = ["Excellent", "Good", "Fair", "Poor"]
y_true_tomatoes = [1, 1, 1, 1, 2, 2, 2, 2, 2, 3, 3, 3,
    4, 4, 4, 4]
y_pred_tomatoes = [2, 1, 3, 1, 2, 2, 2, 3, 4, 3, 4, 2,
    4, 4, 3, 4]

# Generate the confusion matrix
cm = confusion_matrix(y_true_tomatoes,
    y_pred_tomatoes)

# Create the confusion matrix display with proper class labels
disp = ConfusionMatrixDisplay(confusion_matrix=cm,
    display_labels=grades)

# Plot the confusion matrix
disp.plot(cmap=plt.cm.Greens)
plt.title("Confusion Matrix for \nTomato Grades
    Classification")
plt.xlabel("Predicted Grade")
plt.ylabel("True Grade")

# Add custom text legend to show the grade mapping
plt.text(4.5, 1.5, "Grade Mapping:", fontsize=10,
    weight='bold', color='black')
plt.text(4.5, 1.3, "1 → Excellent", fontsize=9,
    color='black')
plt.text(4.5, 1.1, "2 → Good", fontsize=9,
    color='black')
plt.text(4.5, 0.9, "3 → Fair", fontsize=9,
    color='black')
plt.text(4.5, 0.7, "4 → Poor", fontsize=9,
    color='black')
```

```
# Show the plot
plt.tight_layout()
plt.show()

ConfusionMatrixDisplay.from_predictions(
    y_true_tomatoes, y_pred_tomatoes)
```

This prints out the following confusion matrix:

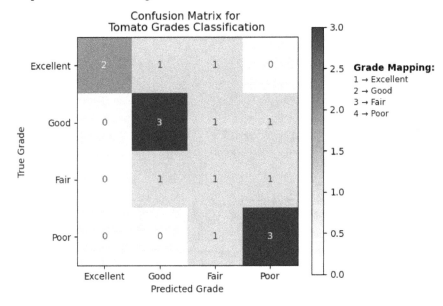

Figure 11.4 – Tomato example confusion matrix

The confusion matrix gives you a visual to see how well each classifier does. In this example, the classifier for Poor only gets one wrong, while the other three classifiers all get two wrong. Additionally, there are more examples of Good tomatoes than the other classifications, and the model gets three of the five correct. For Fair tomatoes, there are only three examples and the model only gets one correct.

2. To print out the classification report from scikit-learn, first, import the classification_ report module, then print it out:

```
from sklearn.metrics import classification_report
print(classification_report(y_true_tomatoes,
    y_pred_tomatoes,target_names=grades))
```

This displays the following output:

```
              precision    recall  f1-score    support
```

Excellent	1.00	0.50	0.67	4
Good	0.60	0.60	0.60	5
Fair	0.25	0.33	0.29	3
Poor	0.60	0.75	0.67	4
accuracy			0.56	16
macro avg	0.61	0.55	0.55	16
weighted avg	0.63	0.56	0.57	16

Here, you see the precision, recall, and F_1 score for each classifier, as well as how many examples it has (support), which is used in the weighting. As you may have expected, the F_1 score for the classifier that grades `Poor` tomatoes and only had one wrong is the best at `0.67`. The classifier that grades `Excellent` is also high because its precision was high. Each time it said a tomato was `Excellent`, it was correct. Now, let's look at how the overall scores are calculated.

The overall accuracy for this OVR multi-class classifier is given by the micro average, dividing the sum of the diagonal cells in the confusion matrix by the sum of all the cells. In this case, that is *(2+3+1+3)/16 = 9/16 = 0.56*.

Then the report gives the weighted and macro average, precision, recall, and F_1 scores. The weighted average is calculated by multiplying the score for the class by the size of the class, then dividing by the total number of samples. The macro average is calculated by taking the arithmetic mean across all the classes. These equations are given in *Table 11.6*. To see how these work, let's do an example using precision:

Precision weighted average = ((excellent score × number of excellent items) + (good score × number of good items) + (fair score × number of fair items) + (poor score × number of poor items))/total number of items

$$= ((1 * 4) + (0.6 * 5) + (0.25 * 3) + (0.6 * 4)) / 16$$

$$= (4 + 3 + 0.75 + 2.4) / 16$$

$$= 10.15 / 16$$

$$= 0.63$$

Precision macro average = (excellent score + good score + fair score + poor score)/total number of items

$$= (1 + 0.6 + 0.25 + 0.6) / 4$$

$$= 2.45 / 4$$

$$= 0.61$$

Now you are familiar with how to calculate metrics for when there are multiple classifications. Next, we'll review metrics for regression.

Metrics for regression RMSE R²

In this section, we will revisit two key metrics used to evaluate regression models: **Root Mean Square Error** (**RMSE**) and **R-squared** (**R²**). You were first introduced to these metrics in *Chapter 5*, where we used them to compare the performance of XGBoost to other decision tree models. Let's briefly recap what these metrics represent and how they can guide model evaluation:

- **RMSE**: RMSE measures the average difference between the predicted values and the actual values. It provides an indication of how far off the model's predictions are from the true values on average. RMSE values range from 0 to infinity, where lower values indicate better model performance. A lower RMSE means that the model's predictions are closer to the actual outcomes, making it an important metric for assessing accuracy:

$$RMSE = \sqrt{\left(\frac{\sum_{i=1}^{N}(y_i - \hat{y}_i)^2}{N}\right)}$$

- **R²**: R² is a statistical measure that represents the proportion of the variance in the target variable that is explained by the model. In other words, it shows how well the model's predictions match the actual data. R² values range from 0 to 1, with higher values indicating better performance. An R² value of 1 means that the model perfectly explains the variance in the target variable, while a value of 0 means the model explains none of the variance. Sometimes, R² is expressed as a percentage; for example, an R² of 0.5 could be written as 50%, meaning the model explains 50% of the variance in the target:

$$R^2 = 1 - \frac{\sum_i (y_i - \hat{y}_i)^2}{\sum_i (y_i - \bar{y}_i)^2}$$

Now that you understand the key concepts of RMSE and R², we'll summarize these metrics in a table below that you can easily refer back to as you evaluate model performance throughout the book.

Evaluating how well a model fits the data summary

To summarize the various metrics used to evaluate model performance, the following table outlines the key metrics and what they tell you about the model's fit. These metrics apply across different model types, such as binary classification, multi-class classification, and regression.

Model type	Fit parameter	Equation	What it tells you
Binary classification	Accuracy	$\frac{(TP + TN)}{(TP + FP + TN + FN)}$	How many of the predictions made by the model were correct
	Precision	$\frac{TP}{(TP + FP)}$	How many of the positive predictions made by the model were correct
	Recall	$\frac{TP}{(TP + FN)}$	How many correct positive predictions were made by the model as a fraction of all the positive cases
	F_1 Score	$2 \cdot \frac{(precision \cdot recall)}{(precision + recall)}$	Combination of precision and recall into a single metric
	ROC AUC	Area under the ROC curve	How well the model differentiates between classes
Multi-class classification	Macro precision	$\frac{\sum_{i=1}^{number\,of\,classes} precision_i}{number\,of\,classes}$	Arithmetic mean of all the precision scores
	Weighted precision	$\frac{\sum_{i=1}^{number\,of\,classes} (precision_i)(class\,size_i)}{total\,samples}$	Weighted average of class-wise precision values weighted by class size
	Macro recall	$\frac{\sum_{i=1}^{number\,of\,classes} recall_i}{number\,of\,classes}$	Arithmetic mean of all class-wise recall values
	Weighted recall	$\frac{\sum_{i=1}^{number\,of\,classes} (recall_i)(class\,size_i)}{total\,samples}$	Weighted average of class-wise recall values weighted by class size
	Macro F_1 score	$\frac{\sum_{i=1}^{number\,of\,classes} F_1}{number\,of\,classes}$	Arithmetic mean of all class-wise F_1 score values
	Weighted F_1 score	$\frac{\sum_{i=1}^{number\,of\,classes} (F_1)(class\,size_i)}{total\,samples}$	Weighted average of class-wise F_1 score values weighted by class size
	Macro ROC AUC	$\frac{\sum_{i=1}^{number\,of\,classes} ROC\,AUC_i}{number\,of\,classes}$	Arithmetic mean of all class-wise ROC AUC score values
	Weighted ROC AUC	$\frac{\sum_{i=1}^{number\,of\,classes} (ROC\,AUC_i)(class\,size_i)}{total\,samples}$	Weighted average of class-wise ROC AUC score values weighted by class size

Model type	Fit parameter	Equation	What it tells you
Regression	RMSE	$RMSE = \sqrt{\left(\frac{\sum_{i=1}^{N}(y_i - \hat{y}_i)^2}{N}\right)}$	Measures the difference between the model and the actual values
	R^2	$R^2 = 1 - \frac{\sum_i (y_i - \hat{y}_i)^2}{\sum_i (y_i - \bar{y}_i)^2}$	How much of the variance of the target is predicted by the model

Table 11.6 – Summary of metrics to measure model fit

In addition to the metrics for how well the model fits the data, you will also want to look at performance metrics to assess how well the model is doing and whether you want to adjust hyperparameter settings. Next, you'll learn about measuring training and inference time.

Model performance, training, and inference time

In addition to measuring how well your model fits the data, you also want to ensure that you are producing an efficient model in terms of the required training and inference time. Based on the need for precision, recall, and F_1 score, you may choose a model that takes a bit longer to train because it provides a better result. On the other hand, you may find that you are in a situation where there are diminishing returns, and a model that requires more training time produces only a small improvement in the performance metrics.

To measure the training and inference time required, you can use %%time within a Jupyter notebook as you did in *Chapter 5*. In the Python code, you can use the time.perf_counter() function to measure the execution time of a section of code. In this next example, you'll measure the time to count to 10 using a for loop. To measure training time, replace the for loop with your code to train the model. To measure inference time, replace the for loop with your code to make a prediction.

Measure the time it takes to run a section of Python code. You can use the time.perf_counter() function. Before the block of code to measure, store the start time by calling start_time = time. perf_counter(). After the block of code to measure, store the end time by calling end_time = time.perf_counter(). The time it takes for the code to run is given by the difference between the end and start values. The time.perf_counter() function measures elapsed time in seconds:

```
import time
start_time = time.perf_counter()

# replace this for loop with what you want to measure
for i in range(10):
    print(i)
end_time = time.perf_counter()
elapsed_time = end_time - start_time
print (f"The code executed in {elapsed_time:.5f} seconds")
```

This results in the following output:

```
0
1
2
3
4
5
6
7
8
9
The code executed in 0.00025 seconds
```

You now have all the measurements you need to compare models in terms of their performance (model fit metrics) and efficiency (training and inference time). Next, you will learn about how to tune the parameters in XGBoost to improve these metrics.

Tuning XGBoost hyperparameters to improve model fit and efficiency

In the previous section, we discussed various metrics to measure how well a model fits the dataset. Now, we will explore how to optimize an XGBoost model by tuning its hyperparameters to improve both accuracy and efficiency. Hyperparameter tuning is crucial in extracting the best performance from XGBoost, as it helps balance model complexity and overfitting, while also improving computation efficiency.

What is hyperparameter tuning?

Hyperparameter tuning is an optimization process that involves adjusting parameters that control the training process of the model. These parameters are not learned from the data but are set before the training process begins. By fine-tuning them, you can significantly improve the model's performance.

There are several methods you can use to systematically tune these hyperparameters:

- **Grid search**: This method tries every combination of hyperparameters in a specified range. It is computationally expensive but guarantees finding the best combination within the search space.

- **Random search**: This method samples a random subset of the hyperparameter space, making it faster than grid search, though it may not always find the optimal solution.

- **Bayesian optimization**: This method builds a probabilistic model of the objective function and selects the most promising hyperparameters to try. It is more efficient than grid search and random search for large search spaces.

Before diving into the tuning process, it's important to select the right evaluation metric based on the task at hand. Here are some examples:

- For regression models, RMSE or R^2 might be appropriate
- For binary classification, AUC or F_1 score could be better suited

As you tune hyperparameters, you should evaluate model performance using a training dataset and then check performance on a test dataset to avoid overfitting. Let's now take a deeper dive into the specific hyperparameters that you can tune in XGBoost.

Key XGBoost hyperparameters

Here are some of the key XGBoost hyperparameters:

- **Learning rate** (`learning_rate`): The learning rate controls how much the model adjusts with each update. A smaller learning rate makes the model more cautious but requires more boosting rounds (trees) to converge.

  Tuning tip: Start with a default value of `0.1` and adjust it down (e.g., to `0.01` or `0.001`) if overfitting occurs. Lower learning rates often require increasing the number of trees (`n_estimators`) to maintain accuracy.

- **Number of trees** (`n_estimators`): This parameter specifies the number of boosting rounds (or trees). Too many trees can cause overfitting, while too few might result in underfitting.

  Tuning tip: Start with a modest number (e.g., 100) and increase it gradually while watching for signs of overfitting. You can compensate for too many trees by decreasing the learning rate.

- **Maximum depth of trees** (`max_depth`): This controls the depth of each individual tree. Deeper trees allow the model to learn more complex patterns, but they also increase the risk of overfitting.

  Tuning tip: Typically, values between 3 and 10 work well, but this varies with the dataset. Shallow trees (e.g., `max_depth = 3`) prevent overfitting but might underfit, while deeper trees (e.g., `max_depth = 10`) capture more complexity.

- **Minimum child weight** (`min_child_weight`): This parameter sets the minimum sum of instance weights (Hessian) needed in a child node. It controls the model's sensitivity to outliers by preventing the model from learning overly specific patterns.

  Tuning tip: Increase `min_child_weight` to prevent overfitting if your model performs poorly on validation data. Default values range from 1 to 6, depending on the dataset size.

- **Gamma** (`gamma`): Gamma specifies the minimum loss reduction required to make a split. A higher gamma value makes the algorithm more conservative in splitting nodes, reducing overfitting.

  Tuning Tip: Start with the default of 0 and gradually increase (e.g., to 0.1 or 0.2) if you notice the model is overfitting. Higher values might reduce model complexity.

- **Subsample** (`subsample`): This parameter controls the proportion of training data used for each boosting round. Using less than 1.0 can help prevent overfitting.

  Tuning tip: A common range for subsample is between 0.5 and 1.0. Lowering this value (e.g., to 0.7) forces the model to use only a portion of the data in each round, reducing variance but potentially increasing bias.

- **Columns to be sampled** (`colsample_bytree`): This parameter controls the fraction of features (columns) to be randomly sampled for each tree. Reducing the number of features per tree can help prevent overfitting.

  Tuning tip: Typically, values between 0.5 and 1.0 are used. Lower values reduce overfitting but may miss important relationships between features.

- **Regularization parameters** (`lambda` and `alpha`): These parameters control L2 (ridge) and L1 (lasso) regularization, respectively. They add a penalty for larger weights, which helps control model complexity.

  Tuning tip: Increasing `lambda` (L2) or `alpha` (L1) can reduce overfitting. Start with `lambda` = 1 and `alpha` = 0 and increase as necessary based on cross-validation performance.

Now that we know the key hyperparameters, let's look at how to tune the model.

Practical steps for tuning

Here's a step-by-step approach to tuning XGBoost hyperparameters:

1. **Start with** `learning_rate`: Set a value (e.g., 0.1), and later refine it if the model starts overfitting. Small learning rates require more boosting rounds.

2. **Tune** `n_estimators`: Begin with a small number of trees (e.g., 100), then increase based on performance, adjusting `learning_rate` accordingly.

3. **Adjust tree complexity**: Tune `max_depth`, `min_child_weight`, and `gamma` to control the complexity of each tree. Start with moderate values (e.g., `max_depth` = 6, `min_child_weight` = 1, and `gamma` = 0) and increase or decrease as necessary.

4. **Refine subsampling parameters**: Fine-tune `subsample` and `colsample_bytree` to prevent overfitting. Use values that are less than 1.0 to introduce randomness into the training process.

5. **Regularization**: Adjust `lambda` and `alpha` to control model complexity and prevent overfitting.

6. **Iterate and refine**: Once the core parameters are optimized, go back and fine-tune `learning_rate` again if necessary to balance convergence speed and accuracy.

Hyperparameter tuning is essential to maximizing the performance of your XGBoost models. The settings of these hyperparameters depend heavily on the specific dataset and problem you are solving. By following a structured tuning approach, using a combination of search techniques, and regularly validating your model, you can achieve a well-balanced model that performs well without overfitting.

Summary

In this chapter, we explored the critical steps of evaluating a model's performance and fine-tuning its parameters to enhance accuracy and efficiency. You gained practical experience with both the scikit-learn and XGBoost Python APIs, applying key performance metrics to classification and regression models.

We delved into hyperparameter tuning, learning how to adjust parameters such as learning rate, number of trees, and regularization factors to improve model fit while avoiding overfitting or underfitting. By closely monitoring performance on both training and test datasets, you ensured that your model generalizes well beyond the initial dataset.

In the next chapter, we'll shift focus to *managing the feature engineering pipeline*, a vital component in maintaining consistency and efficiency when deploying models in production environments.

12

Managing a Feature Engineering Pipeline in Training and Inference

Feature engineering is one of the most pivotal steps in any machine learning project, as it transforms raw data into a structured form that can be effectively used by machine learning models. This process involves creating new features, selecting relevant ones, and manipulating data in a way that enhances the predictive power of the model. The quality of feature engineering can make or break a model's performance, and it often requires a deep understanding of both the data and the domain.

In production environments, *consistency* between feature engineering during the training and inference stages is critical. Any misalignment in how features are engineered across these two stages can lead to significant performance degradation. For instance, if the model is trained with one set of transformations or feature logic but a different approach is applied when making predictions, the model's accuracy, reliability, and overall performance will suffer. Therefore, ensuring that feature engineering is seamlessly integrated into both training and inference pipelines is essential for a robust deployment.

In this chapter, you will delve into the intricacies of managing feature engineering pipelines specifically tailored for **time series forecasting** and **regression tasks** using the **XGBoost model**, expanding on the concepts and code examples discussed in *Chapters 7* and *9*.

You will begin by exploring strategies to streamline the feature engineering process. This involves automating repetitive tasks, selecting appropriate transformations, and ensuring that these transformations can be efficiently applied to large datasets. A special focus will be placed on creating reusable, well-structured pipelines that can be deployed across different stages of a project—from initial model training to real-time inference. You will explore tools such as **scikit-learn's Pipeline**, custom functions, and utility scripts, demonstrating how to build robust, end-to-end workflows that simplify deployment. By designing modular and reusable components, we can avoid redundancy and potential errors while ensuring that feature engineering logic is not duplicated or altered across environments.

In this chapter, we will cover the following topics:

- Using pipelines to avoid data leaks

- Introducing scikit-learn's Pipeline

- Implementing a data pipeline in Python

Technical requirements

The code presented in this chapter is available in our GitHub repository: `https://github.com/PacktPublishing/XGBoost-for-Regression-Predictive-Modeling-and-Time-Series-Analysis/blob/main/ch12/feature_engineering_pipeline_train_inference.ipynb`

As usual, you will need to install the software and Python packages in the following list in order to follow along with the chapter:

- Python 3.9 (a virtual environment is recommended)

- pandas 1.4.2 or 2.1.4

- Jupyter Notebook or JupyterLab

- scikit-learn 1.4.2

- XGBoost 2.0.3

Using pipelines to avoid data leaks

One of the most common issues when working with machine learning models, particularly with time series data, is the potential for **data leakage**. Data leakage occurs when information from outside the training dataset is inadvertently used to train the model, often leading to over-optimistic performance metrics. Pipelines can play a vital role in mitigating this risk by ensuring that the same preprocessing steps applied during training are also applied consistently during inference.

For example, when working with time series data, applying transformations such as scaling or encoding to the entire dataset before splitting into training and testing sets can cause future information to leak into the training process. By using a pipeline, you can avoid this issue by ensuring transformations are applied only to the training data during training and then to the test data during inference.

Introducing scikit-learn's Pipeline

Scikit-learn's Pipeline module provides a powerful way to streamline machine learning workflows by chaining data transformation and model training steps into a single, unified process. Instead of manually tracking and applying each preprocessing step, you can package them together into a pipeline, ensuring that all steps are applied consistently and in the correct order.

One of the major advantages of using a pipeline is that it integrates seamlessly with scikit-learn's other tools, such as cross-validation, grid search, and feature selection. This helps to eliminate errors and ensures that the same transformations are applied across both the training and testing stages.

In the following sections, we'll discuss how to manage feature engineering pipelines specifically for time series forecasting and regression tasks using the XGBoost model. Time series data introduces unique challenges—such as the temporal dependence between observations and the necessity to create lagged features, rolling statistics, and other time-based transformations—that require careful attention. On the other hand, regression tasks can vary in complexity depending on the nature of the data and the relationships between features.

Implementing a data pipeline in Python

Now that we've discussed the importance of consistency in feature engineering across training and inference, it's time to dive into the practical side of building data pipelines. In this chapter, we will focus on implementing feature engineering pipelines in Python that will allow you to automate the process, making it repeatable and reliable for both time series forecasting and regression tasks. This reduces the risk of data leakage and ensures that your model performs optimally when it's deployed in production. Let's explore how to implement these pipelines using Python's scikit-learn and XGBoost libraries.

Time series feature engineering in a pipeline

As you learned in *Chapter 9*, time series data poses unique challenges due to the temporal dependencies between observations. In this section, you will perform feature engineering for time series data and use a pipeline to combine the features with model training. Using a pipeline allows you to perform the same tasks again and again on training and test data.

Using a pipeline with time series data

One of the most common feature engineering techniques for time series is the creation of **lagged features**, which allow the model to capture trends and dependencies over time. In this section, you'll practice feature engineering and use a pipeline to fit an XGBoost model and make predictions. Let's get started:

1. First, import the necessary libraries. You have used most of these in previous chapters. A new library is `Pipeline`:

    ```
    import pandas as pd
    import numpy as np
    from xgboost import XGBRegressor
    from sklearn.pipeline import Pipeline
    from sklearn.model_selection import train_test_split
    from sklearn.preprocessing import OneHotEncoder, StandardScaler
    from sklearn.compose import ColumnTransformer
    ```

```
from sklearn.metrics import mean_squared_error, r2_score
from sklearn.impute import SimpleImputer
from sklearn.base import BaseEstimator, TransformerMixin
```

2. Next, create a custom transformer to make lagged features. This uses the concepts introduced in *Chapter 9*. Here, you will create a class called `LagFeatureTransformer` to make lagged features by shifting the values of the time series by a specified number of lags. These lagged values are then used as features in the model:

```python
# Custom transformer for creating lagged features
class LagFeatureTransformer(BaseEstimator,
    TransformerMixin):
    def __init__(self, lags=3):
        self.lags = lags
    def fit(self, X, y=None):
        return self

    def transform(self, X, y=None):
        df = pd.DataFrame(X.copy())
        for lag in range(1, self.lags + 1):
            df[f'lag_{lag}'] = df['Value'].shift(lag)
            df.dropna(inplace=True)
        return df
```

3. Prepare the time series dataset by creating a synthetic time series dataset and apply `LagFeatureTransformer` to generate lagged features for training a regression model:

```python
date_range = pd.date_range(start='1/1/2020',
    periods=100, freq='D')
data = pd.DataFrame({'Date': date_range,
    'Value': np.random.randn(100).cumsum()})
data.set_index('Date', inplace=True)


lagged_data = LagFeatureTransformer(
    lags=3).transform(data)
X = lagged_data.drop(columns=['Value'])
# Features: lagged values
y = lagged_data['Value']
# Target: original values shifted by lag
```

4. Now that you have created the lagged features, you are ready to use them inside a scikit-learn pipeline that includes both feature scaling and model training using XGBoost. The pipeline allows for streamlined preprocessing and model training, ensuring that the same steps are applied consistently across both the training and inference phases:

```
X_train, X_test, y_train, y_test = train_test_split(X,
    y, test_size=0.2, shuffle=False, random_state=42)

# Define the pipeline with scaling and XGBoost model
pipeline = Pipeline(steps=[
    ('scaler', StandardScaler()),  # Feature scaling
    ('model',XGBRegressor(
        objective='reg:squarederror'))
    # XGBoost regression model
])

# Train the pipeline
pipeline.fit(X_train, y_train)

# Make predictions
y_pred = pipeline.predict(X_test)
```

The pipeline you created in this section performs feature scaling and specifies the model type and hyperparameters you wish to use. The steps in the pipeline are performed in the order you set each time the pipeline is called for either model fitting or model prediction tasks. This allows you to repeat a flow of steps (here, just two steps, and you can see how you could expand it to many steps) every time the model is trained or a prediction is made.

Evaluating the time series model

After training, you can evaluate the model's performance using standard regression metrics such as **mean squared error** (**MSE**) and R^2. Let's do that now:

1. Evaluate the model using MSE and R^2 metrics. These metrics are built into scikit-learn in the `metrics` module:

```
mse = mean_squared_error(y_test, y_pred)
r_squared = r2_score(y_test, y_pred)
print(f'Mean Squared Error: {mse}')
print(f'R squared: {r_squared}')
```

This outputs the following:

```
Mean Squared Error: 5.384843571846216
R squared: -2.35336221354622
```

```
                Actual    Predicted
    Date
    2020-03-21 -19.313636 -20.451332
    2020-03-22 -19.821886 -20.661566
    2020-03-23 -18.626756 -21.220581
    2020-03-24 -18.911709 -20.336151
    2020-03-25 -18.478816 -20.611525
```

In this section, you created a pipeline to combine multiple steps for preparing and executing model training for time series data. This pipeline handles feature engineering and model training in a unified process, making it easier to deploy and manage models. Next, you will build a pipeline for a regression model.

Regression feature engineering in a pipeline

As you learned in *Chapter 7*, regression tasks typically involve datasets with numerical and categorical features. Managing missing values and standardizing features are necessary for building effective regression models. Frequently, a number of feature engineering tasks are necessary to prepare a dataset for modeling. Having all those tasks grouped together in a pipeline makes the data preparation process more repeatable. Next, you will perform all the same feature engineering steps you did in *Chapter 7* in a pipeline.

Using a pipeline with regression data

Real-world datasets often contain missing values in both numeric and categorical variables. In this section, you will build a pipeline that performs the feature engineering tasks needed to prepare the housing dataset from *Chapter 7* for modeling, train a model with XGBoost, and perform inference.

Handling missing values in categorical data

To handle missing values in categorical data, you can apply different strategies depending on the proportion of missing entries. In this example, we will use a custom transformer called `MissingValueImputer`, which either replaces missing values with the string `Missing` for variables with a high proportion of missing values or imputes the most frequent category for variables with fewer missing entries. This ensures that the dataset is complete before being fed into the model.

`BaseEstimator` and `TransformerMixin` from scikit-learn are essential for creating reusable and consistent transformers that integrate seamlessly into scikit-learn's ecosystem. These classes allow us to create custom transformers that can be incorporated into pipelines. `BaseEstimator` provides methods such as `get_params` and `set_params`, making your custom transformer compatible with pipelines and hyperparameter tuning, while `TransformerMixin` adds the `fit_transform` method, ensuring the transformer can be used in both training and inference stages.

Here's the custom transformer:

```python
class MissingValueImputer(BaseEstimator, TransformerMixin):
    def __init__(
    self, for_missing_string, for_frequent_category):
        self.for_missing_string = for_missing_string
        self.for_frequent_category = for_frequent_category
        self.frequent_categories = {}

    def fit(self, X, y=None):
        # Store the most frequent category for variables
        #with few missing observations
            for var in self.for_frequent_category:
                self.frequent_categories[var] =X[
                    var].mode()[0]
            return self

    def transform(self, X, y=None):
        # Replace missing values with "Missing" for
        #specific variables
        X[self.for_missing_string] = X[
            self.for_missing_string].fillna('Missing')
            # Replace missing values with the most frequent
            #category for specific variables
            for var in self.for_frequent_category:
                X[var] = X[var].fillna(
                    self.frequent_categories[var])
            return X
```

This transformer ensures that categorical variables are properly preprocessed before model training, either by imputing missing values with the most frequent category or replacing them with the string `Missing`.

By inheriting from `BaseEstimator` and `TransformerMixin`, your transformer becomes reusable and compatible with scikit-learn's entire pipeline framework, enabling efficient, consistent transformations during both training and inference.

Preparing the dataset

Next, we will prepare the dataset for modeling. We use the housing dataset from *Chapter 7*, which contains both numeric and categorical features. The target variable is `SalePrice`, and we will preprocess both feature types accordingly:

```python
data = pd.read_csv("house_pricing.csv")
```

```
# Define feature columns
numeric_features = ['LotArea', 'YearBuilt', '1stFlrSF',
    '2ndFlrSF', 'GrLivArea']
categorical_features = ['Neighborhood', 'HouseStyle',
    'GarageType', 'SaleCondition']

# Handle missing categorical variables
cat_vars_with_na = [var for var in categorical_features if data[var].
isnull().sum() > 0]
for_missing_string = [var for var in cat_vars_with_na if data[var].
isnull().mean() > 0.1]
for_frequent_category = [var for var in cat_vars_with_na if data[var].
isnull().mean() < 0.1]
# Target column
target = 'SalePrice'
```

Next, split the data into features and `target` and perform the `train/test` split:

```
X = data.drop(columns=[target, 'Id'])
y = data[target]
X_train, X_test, y_train, y_test = train_test_split(X, y,
    test_size=0.2, random_state=42)
```

Now the data is prepared, you can move on to building the pipeline.

Building the pipeline

Now that we've prepared the dataset, we can build the full pipeline that combines both numeric and categorical feature preprocessing with model training using XGBoost:

```
numeric_transformer = Pipeline(steps=[
    ('imputer', SimpleImputer(strategy='mean')),
     # Handle missing values
    ('scaler', StandardScaler())      # Scale features
])

categorical_transformer = Pipeline(steps=[
    ('imputer', SimpleImputer(strategy='most_frequent')),
    # Handle missing categorical values
    ('onehot', OneHotEncoder(handle_unknown='ignore'))
     # One-hot encode categorical features
])

preprocessor = ColumnTransformer(
    transformers=[
        ('num', numeric_transformer, numeric_features),
        ('cat', categorical_transformer,
```

```
              categorical_features)
    ])

pipeline = Pipeline(steps=[
    ('missing_imputer', MissingValueImputer(
        for_missing_string, for_frequent_category)),
    # Custom missing value handling
    ('preprocessor', preprocessor),
    # Preprocessing step
    ('model', XGBRegressor(objective='reg:squarederror'))
    # XGBoost regression model
])
```

Evaluating the model

Now that the pipeline is set up, you can train and evaluate the regression model using the MSE and R^2 metrics:

```
# Train the pipeline
pipeline.fit(X_train, y_train)
# Make predictions
y_pred = pipeline.predict(X_test)
# Evaluate the model
mse = mean_squared_error(y_test, y_pred)
print(f'Mean Squared Error: {mse}')

# Display the predicted vs actual values
result = pd.DataFrame({'Actual': y_test,
    'Predicted': y_pred}, index=X_test.index)
print(result.head())
```

This prints the following output:

```
Mean Squared Error: 1130202899.0655065
       Actual      Predicted
892    154500   142317.156250
1105   325000   364155.031250
413    115000    93371.398438
522    159000   135780.375000
1036   315500   266444.375000
```

In this section, we combined data preparation through feature engineering and model training using pipelines. By using scikit-learn's BaseEstimator and TransformerMixin classes, we ensured that the code is reusable and compatible with scikit-learn pipelines, allowing efficient and consistent transformations during both training and inference.

Summary

In this chapter, you learned how to effectively manage feature engineering pipelines for both time series forecasting and regression tasks using the powerful tools provided by scikit-learn's Pipeline API. As you've seen, the ability to package preprocessing steps into a unified workflow is crucial for ensuring consistency and reproducibility across both training and inference stages.

For time series forecasting, you explored how to create lagged features, which allow you to capture the temporal dependencies in your data. The custom transformer you built for generating lag features makes it easy to handle shifting time steps, enabling your model to learn from past values. You now understand how to integrate this feature engineering step into a pipeline that can preprocess the data, scale it, and feed it into an XGBoost model, making the entire process seamless and reusable.

When dealing with regression tasks, you faced a different set of challenges, especially in handling missing values in categorical features. By learning how to build a custom transformer to replace missing values with meaningful substitutes—either the string `Missing` for high missing rates or the most frequent category for lower rates—you have developed a flexible approach that ensures your model gets clean and usable data. You also applied a variety of preprocessing techniques, from scaling numerical features to one-hot encoding categorical features, creating a comprehensive pipeline that combines both numerical and categorical feature engineering.

Through these examples, you've mastered the ability to build a full-featured pipeline that can not only handle diverse types of data but also ensure that the same transformations are applied during both training and inference. This consistency is critical when deploying models in production because it reduces the risk of introducing errors or inconsistencies between the training data and real-world data the model will eventually face.

Incorporating these pipelines into your workflow will improve the scalability of your machine learning projects. The tools you now have in your toolkit allow you to easily maintain and update models, saving time and ensuring high-quality, reproducible results. Furthermore, you've gained a deeper understanding of how to manage feature engineering in a structured and systematic way, which is essential when handling complex data environments.

By embracing the principles outlined in this chapter, you are now equipped to handle both time series forecasting and regression tasks more efficiently, enabling you to build models that generalize well and perform reliably in production. As you continue to explore and implement machine learning models, these techniques will serve as the foundation for handling the feature engineering process with confidence and precision.

In the next chapter, you will learn about how to deploy your XGBoost model into production. You will also learn about using multithreading and distributed computing to handle large datasets with XGBoost.

13

Deploying Your XGBoost Model

In this chapter, you'll explore how to deploy your XGBoost model into production environments and utilize it for real-time inference. You'll learn about key aspects of model maintenance, including monitoring performance with metrics, which you learned about in *Chapter 11*. You'll also see when and how to retrain models, as well as explore the multithreaded and distributed computing options for optimizing training speed. Lastly, you'll dive into cloud-based deployment using containerization technologies such as Docker.

Here's a breakdown of the key topics we'll cover:

- Training a model using multithreaded and distributed computing with XGBoost
- Packaging a model for production deployment
- Deploying a model using containers
- Servicing your model using the REST API

Technical requirements

You can continue using the same virtual environment that you set up earlier in this book since it already contains the necessary packages. The code samples for this chapter can be found in this book's GitHub repository: `https://github.com/PacktPublishing/XGBoost-for-Regression-Predictive-Modeling-and-Time-Series-Analysis`.

For this chapter, you'll be using the following tools and libraries:

- Windows Subsystem for Linux (WSL)
- Python 3.9
- XGBoost 1.7.3
- NumPy 1.21.5
- pandas 1.4.2

- scikit-learn 1.4.2

- Seaborn

- Anaconda

- VS Code

- Dask

- Docker

- Flask 3.0

- Gunicorn

Using Linux for Deployment

When deploying your XGBoost model, it's advisable to use **Linux** for your container environment. Linux is the most widely adopted operating system for distributed computing and offers greater flexibility for deployment across multiple cloud providers. If you're developing on a Windows-based machine, you can use **Windows Subsystem for Linux** (**WSL**) to create a Linux environment within your Windows setup.

Training a model using multithreaded and distributed computing with XGBoost

Let's start with a critical aspect of deploying machine learning models: **training time**. As you saw in *Chapter 5*, model training can be a time-consuming process, especially when working with large datasets or when frequent retraining is necessary.

XGBoost offers built-in support for **multithreaded computing**, which can significantly speed up the training process by utilizing multiple CPU cores. In this section, you'll explore how to enable **multithreading** for XGBoost training, which works on both Windows and Linux systems. Then, you'll delve into **distributed computing** options using **Dask** (www.dask.org) on Linux so that you can scale model training across clusters or in cloud environments. Let's begin.

Using XGBoost's multithreaded features

XGBoost has built-in support for multithreaded computing, which allows you to speed up model training by utilizing multiple CPU cores. You can control this by setting the `nthread` parameter, which determines the number of threads to use. By default, XGBoost will automatically use the maximum number of available threads.

It's important to note that if you're using Dask, any value you set for `nthread` within XGBoost will take precedence over Dask's default configuration. The following example demonstrates how the multithreading parameter works. We'll revisit the *California housing dataset* that you worked with in *Chapter 4*:

1. Create a Python file to demonstrate XGBoost's multithreaded functionality. We've started with a header and named the file `multithreaded.py`.

2. Import the necessary modules. You can load the California housing dataset from scikit-learn (`sklearn`). You'll also be using `pandas`, `numpy`, a module called `time` to track how long code execution takes, and, of course, `xgboost`:

```
import pandas as pd
import numpy as np
import time
import xgboost as xgb
from sklearn.metrics import r2_score
from sklearn import datasets
from sklearn.model_selection import train_test_split
```

3. Now, you can load in the California housing dataset and perform the train-test split using scikit-learn, as you did previously:

```
housingX, housingy = datasets.fetch_california_housing(
    return_X_y=True, as_frame=True)

X_train, X_test, y_train, y_test = train_test_split(
    housingX,housingy, test_size=0.2, random_state=17)
```

4. Previously, you used the scikit-learn interface for XGBoost. In this example, you'll use the XGBoost API for Python. One difference is that XGBoost uses a data structure called a **DMatrix** to manipulate data. So, the first thing you need to do is convert the dataset from `numpy` or `pandas` form into `Dmatrix` form by using the `DMatrix` function and passing in the data and the labels. In this case, we'll be using `dtrain = xgb.DMatrix(X_train, y_train)` for the training dataset; do the same for the test dataset:

```
dtrain = xgb.DMatrix(X_train, y_train)
dtest = xgb.DMatrix(X_test, y_test)
```

Now, the data is in a format that XGBoost can manipulate with efficiency. As mentioned in *Chapter 3*, XGBoost does some sorting and performs other operations on the dataset to speed up execution.

5. At this point, you're ready to train a model using the XGBoost API and the multithreading feature. By default, XGBoost uses the maximum number of threads available. To see the difference, train the model with just two threads, and then increase the maximum number of logical processors you have in your computer. You'll need to use the time module to get the computation time and print it out so that you can compare the results. First, save the start time with the following line of code:

    ```
    train_start = time.time()
    ```

6. You can set the training parameters for XGBoost by creating a dictionary with the parameters as key-value pairs. You can configure all the parameters listed in the *Hyperparameters* section of *Chapter 5*. Here, set eta = 0.3 (the learning rate), booster = gbtree, and nthread = 2:

    ```
    param = {"eta": 0.3, "booster": "gbtree",
        "nthread": 2}
    ```

7. Now that the training parameters have been set, you can train the model and save the end of the execution time by using the following code:

    ```
    housevalue_xgb = xgb.train(param, dtrain)
    train_end = time.time()
    ```

8. Print the execution time with a formatted print statement while subtracting train_start from train_end and converting it into milliseconds by multiplying by 10^3:

    ```
    print ("Training time with 2 threads is :{
        0:.3f}".format((train_end - train_start) * 10**3),
        "ms")
    ```

9. Now, repeat the code and increase the number of threads XGBoost uses by changing the value of nthread. Since our computer has eight logical processors, I've chosen 8:

    ```
    train_start = time.time()
    param = {"eta": 0.3, "booster": "gbtree",
        "nthread": 8}
    housevalue_xgb = xgb.train(param, dtrain)
    train_end = time.time()

    print ("Training time with 8 threads is :{
        0:.3f}".format((train_end - train_start) * 10**3),
        "ms")
    ```

10. To ensure the model is working as expected, you can make a prediction and check the R^2 value. You can also time the prediction. To make a prediction with the Python API, just call the `predict` method on your model and pass the test dataset:

```
pred_start = time.time()
ypred = housevalue_xgb.predict(dtest)
pred_end = time.time()

print ("Prediction time is :{0:.3f}".format((
    pred_end - pred_start) * 10**3), "ms")

xgb_r2 = r2_score(y_true=y_test, y_pred= ypred)
print ("XGBoost Rsquared is {0:.2f}".format(xgb_r2))
```

11. Running this script results in the following output. Please note that the execution time on your computer will be different:

```
Training time with 2 threads is :237.088 ms
Training time with 8 threads is :130.723 ms
Prediction time is :2.012 ms XGBoost
Rsquared is 0.76
```

On our computer, going from two to eight threads sped up training by over 44%. This demonstrates the benefit XGBoost provides with multithreading. Recall that by default, it will use the maximum number of threads available. Next, you'll learn about using XGBoost with distributed compute by using Dask on Linux.

Training with distributed compute

So far, you've used XGBoost on a single local computer. Sometimes, though, a dataset is very large, so to speed up execution, you'll want to share the work across a cluster of machines. From this point onwards in this chapter, we'll be using *Linux* since it's the most common operating system for distributed compute and lets you deploy your model with multiple cloud service providers.

As mentioned earlier in this chapter, you can use **Dask** to enable parallel computing. When using Dask, you need to store your data in **Dask DataFrames**. Dask sets up a computation as a task graph and then calls the `compute` method to run it on a scheduler. Schedulers can be single or multithreaded and can be on a single machine or distributed in a cluster. Dask works both on your local machine and with Kubernetes or a cloud service such as AWS or Azure. XGBoost has a Dask feature that you'll be using in this section.

To start using Dask, you'll need to implement an example on your local machine. When you want to move to a Kubernetes cluster or cloud provider such as AWS or Azure, you can comment out the local machine cluster statement and add one for the appropriate service. Details about using the various cloud providers, such as AWS and Azure, are available in the Dask documentation at `https://cloudprovider.dask.org/en/latest/index.html`.

This example from the Dask documentation (`https://docs.dask.org/en/latest/deploying.html`) shows how you can easily change from running Dask on your local computer to running Dask in a Kubernetes cluster (`cluster = KubeCluster()`):

```
# You can swap out LocalCluster for other cluster types
from dask.distributed import LocalCluster
from dask_kubernetes import KubeCluster
# cluster = LocalCluster()
cluster = KubeCluster()   # example, you can swap out for
                          #Kubernetes
client = cluster.get_client()
```

With this out of the way, you're ready to complete a Dask example on your local computer using `LocalCluster()`. Unlike our previous examples, you'll run this one not as a Jupyter notebook but as a Python `.py` file. In addition, you'll use Linux when running the code. First, you need to install `dask` into a virtual environment using `conda`:

1. Open a Linux Terminal window and create a `conda` environment for this project. Then, activate the environment by running the following commands:

    ```
    conda create --name xgboost_book_project
    conda activate xgboost_book_project
    ```

    Replace `xgboost_book_project` with the name of your virtual environment.

2. Then, run the following command to install Dask:

    ```
    conda install dask
    ```

    You may be prompted to install new packages and update others. Type `y` to proceed.

    This will result in the following message:

    ```
    Downloading and Extracting Packages

    Preparing transaction: done
    Verifying transaction: done
    Executing transaction: done
    ```

3. Now, you can create a Python script to test Dask on your local Linux machine. Start a file called `dasktest.py` and import the `pandas`, `numpy`, `time`, and `xgboost` packages, `metrics`, `datasets`, and `train_test_split` from scikit-learn, and various `dask` components. Here's what the top of the script looks like:

    ```
    import pandas as pd
    import numpy as np
    import time
    import xgboost as xgb
    ```

```
from sklearn.metrics import r2_score
from sklearn import datasets
from sklearn.model_selection import train_test_split
from dask import dataframe as dd
from dask.distributed import Client, LocalCluster
from xgboost import dask as dxgb
```

4. Next, you must define a function that makes the Dask calls for training purposes. Pass information about the distributed system to use and the training dataset to the function:

```
def train(client, dtrain):
```

5. The first thing you must do is start a timer so that you can keep track of how long training takes and see the impact of Dask settings such as the number of workers:

```
train_start = time.time()
```

6. Next, perform training using XGBoost. In the case of distributed computing, XGBoost uses the approx tree_method. You can set it directly or use the default, auto, which will switch to approx when you execute the function:

```
housevalue_dask = xgb.dask.train(
    client,
    {"verbosity": 2, "tree_method": "approx",
     "objective": "reg:squarederror"},
    dtrain,
    num_boost_round=4,
    evals=[(dtrain, "train")],
)
```

7. Finish this function by closing out the timer, printing the time, and returning the trained model so that it can be reused later:

```
train_end = time.time()
print("Training time with dask is :{
    0:.3f}".format((
        train_end - train_start) * 10**3), "ms")
return housevalue_dask
```

8. The next function to build is the prediction. You'll want to pass the distributed computing client, the model you trained, and the testing dataset:

```
def predict(client, model, dtest):
```

9. Next, start a timer for the prediction:

```
pred_start = time.time()
```

10. Then, call `xgb.dask.predict` to make the prediction:

```
ypred = xgb.dask.predict(client, model, dtest)
```

11. Finish this function by calling `time` to measure how long it takes and using the Dask functionality in XGBoost to predict a value and return the predicted y values:

```
pred_end = time.time()
print("Prediction time with dask is :{
    0:.3f}".format((pred_end - pred_start) *
    10**3), "ms")


return ypred
{0:.2f}".format(xgb_dask_r2))
```

12. In this example, you can set the number of workers to 4 with `n_workers = 4` and the number of threads per worker (`threads_per_worker`) to 1. Feel free to experiment with changing these settings and see the effects on the training and prediction times:

```
if __name__ =="__main__":
    with LocalCluster(n_workers = 4,
        threads_per_worker = 1) as cluster:
            with Client(cluster) as client:
```

13. Now, you'll use Dask to predict California housing values. To do so, you'll be using the California housing dataset that you used previously. First, start the function and load in the California housing dataset from scikit-learn:

```
# load the California Housing data set from
#scikit-learn
housingX, housingy =
datasets.fetch_california_housing(
    return_X_y=True, as_frame=True)
```

14. Then, perform `train_test_split`, as you did previously:

```
X_train, X_test, y_train, y_test = train_test_split(
    housingX,housingy, test_size=0.2, random_state=17)
```

15. Dask requires values to be in a Dask DataFrame. So, you must use the `from_pandas` method to convert the output from `train_test_split` into Dask DataFrames:

```
dm_X_train = dd.from_pandas(X_train,
    npartitions=1)
dm_X_test = dd.from_pandas(X_test,
    npartitions=1)
dm_y_train = dd.from_pandas(y_train,
```

```
            npartitions=1)
dm_y_test = dd.from_pandas(y_test,
    npartitions=1)
```

16. Then, combine the data with the labels into a `DaskDMatrix` structure for both the train and test data:

```
dtrain= xgb.dask.DaskDMatrix(client,
    dm_X_train, dm_y_train)
dtest= xgb.dask.DaskDMatrix(client,
    dm_X_test, dm_y_test)
```

17. Now, you're ready to call the `train` and `predict` functions that you defined previously:

```
housevalue_dask = train(client, dtrain)
y_pred = predict(client, housevalue_dask,
    dtest)
```

18. Finish the script by calculating the R^2 value for the Dask XGBoost model:

```
xgb_dask_r2 = r2_score(y_true=dm_y_test,
    y_pred= y_pred)
print("XGBoost with dask Rsquared is {
    0:.2f}".format(xgb_dask_r2))
```

19. With that, the script is complete. You can run it by typing the following Linux command in a Terminal:

```
python3 -m dasktest
```

This results in the following output. Recall that the times may vary, depending on how powerful (or not) your computer is:

```
[15:00:50] task [xgboost.dask-0]:tcp://127.0.0.1:33675 got new rank 0
[15:00:50] INFO: /croot/xgboost-split_1675457761144/work/src/data/
simple_dmatrix.cc:102: Generating new Gradient Index.
[0]     train-rmse:1.44755
[1]     train-rmse:1.10970
[2]     train-rmse:0.88376
[3]     train-rmse:0.74196
Training time with dask is :292.709 ms
Prediction time with dask is :63.967 ms
XGBoost with dask Rsquared is 0.55
```

In this section, you used Dask to train and predict an XGBoost model over a cluster of machines, or multiple workers on a single machine. This capability is helpful when a dataset is very large and you wish to speed up execution. Since Dask is flexible, it can run on a local machine, or you can scale your model with Kubernetes or a cloud service such as AWS or Azure. In the next section, you'll learn how to prepare your model so that it can be deployed in a production environment, rather than it having to be run on your local machine.

Packaging a model for production deployment

So far in this chapter, you've been interacting with the models you've developed directly within Python by making calls to functions and methods that train a model or use a model to make a prediction. In a production environment, you won't want to make Python calls directly. You'll either want to create a web application that interfaces with your model, or you'll want to use an **application programming interface (API)**. An API allows a user to access the functions and methods they need to train and use your model without having to know how to code in Python. Building an API allows a user to call your model from their web application. To better understand this, let's look at the parts of a model that's running in a production environment:

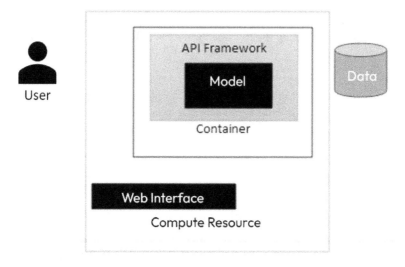

Figure 13.1 – Components of a model running in a production environment

Figure 13.1 shows your model inside an API framework. You'll learn more about this shortly. Both the model and the API framework are inside a container. This container is running on a compute resource that has a web interface. There may also be data storage for your model in a database that can be separate from the compute resource and your model. To access your model, the user makes API calls via the web through the web interface to the API framework running in the container.

But first, let's follow a simpler path to create a web application from a model using Streamlit. Then, later in this chapter, you'll learn how to use the Flask framework to create a **REST API** for your model. REST stands for **Representational State Transfer** and is a standard for how an API should work. After, you'll learn how to use Docker to put your framework and model into a container.

Turning your model into an application with Streamlit

Once you've developed and trained your XGBoost model pipeline, as discussed in *Chapter 12*, the next step is to turn it into an application that allows you to interact with it. **Streamlit**, a Python-based framework, makes this process seamless by providing a simple way to build web applications for machine learning models. With Streamlit, you can deploy your XGBoost model and allow users to input data and make real-time predictions.

Here's a step-by-step approach to turning your model into a web application using Streamlit:

1. Before proceeding, ensure that you have Streamlit installed in your environment. You can install it via `pip`:

    ```
    pip install streamlit
    ```

2. Streamlit makes it easy to build an interface where users can input data and view predictions from your model. In the application, you can use `st.number_input` for numeric features and `st.selectbox` for categorical features. This helps users provide input that matches the model's training data. Here's how you can implement this:

    ```
    import streamlit as st
    import pandas as pd

    # User Input for Predictions
    st.subheader("Make a New Prediction")
    input_data = {}

    # Loop through numeric features and use number_input for numeric
    columns
    for col in numeric_features:
        input_data[col] = st.number_input(f"Enter {col}:",
            value=float(X[col].mean()))
    # Loop through categorical features and use selectbox #for
    categorical columns
    for col in categorical_features:
        unique_values = X[col].unique()
        input_data[col] = st.selectbox(f"Select {col}:",
            options=unique_values)
        input_df = pd.DataFrame([input_data])
        if st.button("Predict"):
            prediction = pipeline.predict(input_df)
            st.write(f"Predicted House Price:
                ${prediction[0]:,.2f}")
    ```

The preceding code is based on the code that was discussed in *Chapters 7, 8*, and *12*. For features such as `LotArea` and `YearBuilt`, the user can input numeric values using the `st.number_input` widget. For categorical features such as `Neighborhood` and `HouseStyle`, users can choose from a predefined list of categories using `st.selectbox`. Once the input has been provided, users can click the **Predict** button, which causes the model to make a prediction.

Your pre-trained XGBoost model has been integrated into this application using the pipeline you built earlier in *Chapter 12*, and is run in `prediction = pipeline.predict(input_df)`. The prediction is based on the user's input data, and the result is displayed on the application.

3. Now, you're ready to launch the application. To run the application on your local machine, save the Streamlit code in a Python file (for example, `streamlit_app.py`) and run the following command in your Terminal:

```
streamlit run streamlit_app.py
```

Once you've executed this command, you'll see the following web page that you can interact with on `localhost:8501`:

Figure 13.2 – Streamlit web application for the housing prices model

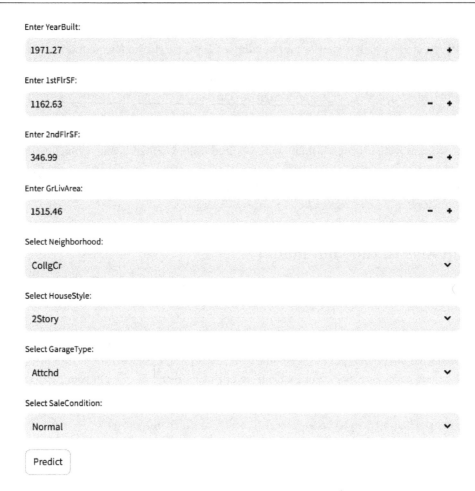

Figure 13.3 – Streamlit application data entry fields

Deploying the application

Once your Streamlit application is ready, it can be deployed on a local machine for internal testing or hosted on cloud platforms for broader access. Services such as Streamlit Cloud, Heroku, and AWS allow you to host your Streamlit application for external users, enabling a fully interactive, web-based machine learning application.

Now that you've built a web application with Streamlit for your model, you'll learn about using the Flask framework to create a **REST API** for your model. You'll use this same Flask framework inside a Docker container.

FastAPI as an alternative to Flask and Green Unicorn (Gunicorn)

FastAPI is a modern web framework for building APIs with asynchronous support natively. While this chapter focuses on deployment with **Flask** and **Gunicorn**, which together enable scalable and performant applications, FastAPI offers an alternative with native asynchronous capabilities, making it ideal for handling high-concurrency applications.

If you're interested in asynchronous frameworks, FastAPI could be a good choice due to its speed and ease of use with async calls. However, for synchronous model inference tasks, Flask and Gunicorn provide reliable scalability and work seamlessly with traditional Python libraries.

FastAPI won't be covered in detail here, but if you'd like to explore it further, here's a recommended resource: `FastAPI: Modern, Fast (High-performance) web framework for Python` (https://fastapi.tiangolo.com/).

Creating APIs to call your model

As you develop your model, think about the tasks you'll want to perform with it once it's deployed in production. For each of these tasks, you'll need to build a method or function that can be called via the API.

For your model to work in the cloud, you'll need to save both it and your training pipeline so that they can be moved from your computer to the cloud for processing. You can save these using a **pickle** file. This allows you to easily transport your model from one place to another. To put your model into a pickle file, follow these steps:

1. Begin by installing the `pickle` package in your `conda` environment:

    ```
    conda install pickle
    ```

2. In the file where you created the model and the pipeline, import `pickle`:

    ```
    import pickle
    ```

3. Save your pipeline as a pickle file with `pickle.dump`:

    ```
    with open('XBGpipeline.pkl', 'wb') as file:
        pickle.dump(pipeline, file)
    ```

You'll likely need to retrain your model from time to time, either based on changes to the environment in which the model is running or due to degradation in the model's performance. Say, for example, you're creating a classification model, such as the one outlined in our tomato example in *Chapter 11*. When a new classification – perhaps "Supreme" – is introduced into the environment, you need to retrain the model; otherwise, it won't know to place tomatoes into that classification. So, you need to build a method to retrain the model.

Similarly, if any of the fit metrics for your model have degraded, you should retrain it on more recent data. So, another method you could create would check the model fit.

The primary use of the model will also need a method, which is to either classify or predict values based on input data.

This adds up to at least three methods you'll want to create for the following purposes:

1. Retraining your model.
2. Making a prediction.
3. Monitoring model fit metrics.

In this section, you'll use the time series model that you made in *Chapter 9* and build these three functions for it in your Flask environment. Let's start by building a method to retrain the model.

Retraining your model

To retrain the model, you'll need to add any new data you want to use for training to your training dataset. You may have this dataset in storage as a DMatrix, a Flask dataset, or even in a database, depending on your needs. Then, you'll need to call the training function. Finally, you'll save the updated version of the model.

Let's write a method that will retrain the model from *Chapter 9*. This method retrieves the model from a pickle file, refits it on the new data, and saves it back as a pickle file:

```
def retrain_model(model,X_train, y_train):
    #load the model from pickle file, retrain the model and
    #save to pickle
    with open(model, 'rb') as file:
        loaded_model_pickle = pickle.load(file)
    loaded_model_pickle.fit(X_train, y_train)
    base_name, extension = os.path.splitext(file_path)
    new_file_path = f"{base_name}_retrained{extension}"
    with open(new_file_path, 'wb') as file:
        pickle.dump(loaded_model_pickle, file)
```

Now, you have a method that can be called via an API to retrain the time series model. Next, let's make a method that will use the model to make a prediction.

Making a prediction

The second method to create will use the model and either make a prediction or classify data depending on the type of model you've created.

For our time series example, you'll be making a prediction by retrieving the model from the pickle file and then running the `.predict` method. The following code loads the model for each prediction. Depending on your needs, you may wish to load the model once – say, when the user starts an API session – and allow multiple predictions to be made:

```
def predict(model, X_test):
    #use the loaded model to predict a result
    with open(model, 'rb') as file:
        loaded_model_pickle = pickle.load(file)
    predictions_pickle = loaded_model_pickle.predict(
        X_test)
    return predictions_pickle
```

In the next section, you'll build a method to monitor how the model is performing.

Monitoring model fit metrics

To monitor model fit, you'll need to create a function that checks the predicted values against a test dataset, just as you did in *Chapter 11*. The metrics you'll use will depend on the problem your model has been developed to address. For example, if you're doing tomato grading, as you did in *Chapter 11*, you could use a confusion matrix and classification report as your metrics. If you're predicting a value, you might select MSE and/or R^2 as your metrics. In this section, you'll use both to evaluate the time series model:

1. Start by making a class to hold the results of model evaluation – that is, `mse` and `r_squared`:

```
class EvalMetrics:
    def __init__(self):
        self.mse = 0
        self.r_squared = 0
```

2. Next, write a function to evaluate the model. The following code runs the prediction function and then calculates the metrics:

```
def evalmodel(model, X_test, y_test):
    y_pred = predict(model, X_test)
    EvalMetrics.mse = mean_squared_error(y_test,
        y_pred)
    EvalMetrics.r_squared = r2_score(y_test, y_pred)
    return EvalMetrics
```

Here, we've only included the relevant parts of the code. For the complete code, please refer to this book's GitHub repository: `https://github.com/PacktPublishing/XGBoost-for-Regression-Predictive-Modeling-and-Time-Series-Analysis`. In this section, you created a method to monitor the model's performance. This wraps up the minimum methods needed for you to use an API with your model. Next, you'll learn how to use Flask to call these functions with an API.

Creating an API for your model using Flask

To provide API access to your model, you'll need a framework. **Flask** is a lightweight web framework that allows you to easily create an API for your model. With Flask, you can configure endpoints to make predictions, add new data, and retrain your model on fresh datasets. By utilizing the REST API standard, Flask enables seamless communication between your model and external systems through HTTP/HTTPS requests, allowing for real-time inference and model updates via a web interface.

> **Flask versus Django**
>
> While **Django** is another popular framework that offers similar functionality, we're using Flask here due to its lightweight nature and flexibility. Unlike Django, Flask is a **microframework**, meaning it doesn't include unnecessary components by default, giving you the freedom to add only the features you need as your project evolves. This makes Flask ideal for smaller projects or scenarios where you're still determining the full scope of the application.
>
> On the other hand, Django is more suited for *large-scale production applications* that require comprehensive administrative controls, security features, and built-in tools. It offers a more complete set of features out of the box, making it a better choice for complex applications with defined requirements from the outset.
>
> In summary, Flask is perfect for smaller models, rapid prototyping, and projects where you need more flexibility and simplicity. Django is ideal for full-fledged production systems with extensive functionality and robust security requirements. Depending on your project's needs, you can choose the framework that best suits your goals.

To allow you to send and receive commands to your code, Flask will start a web service. To see how this works, you can build a quick web application to display "Hello World!" using Flask's *Quickstart* guide at `https://flask.palletsprojects.com/en/3.0.x/quickstart/`. To understand this process, follow these steps:

1. Install Flask using `conda install`:

    ```
    conda install flask
    ```

    This results in the following output:

    ```
    Retrieving notices: ...working... done
    Collecting package metadata (current_repodata.json): done
    Solving environment: done


    ## Package Plan ##
    ```

```
    environment location: C:\Users\joyce\anaconda3\envs\xgboost_
book_project

  added / updated specs:
    - flask
Downloading and Extracting Packages
Preparing transaction: done
Verifying transaction: done
Executing transaction: done
```

2. Create a new file for the Flask application called `FlaskTest.py`:

    ```
    vi FlastTest.py
    ```

3. Inside `FlaskTest.py`, import the `Flask` class. Using `Flask` with a capital letter indicates a class:

    ```
    from flask import Flask
    ```

4. Next, create an application by making an instance of the `Flask` class and passing it the `__name__` magic method. Flask uses `__name__` to know where the Python file that's calling it is located, which, in turn, tells Flask where to find the resources and files it needs to run the application:

    ```
    app=Flask(__name__)
    ```

5. Tell Flask which URL to respond to with `route()`. You can have it respond to all traffic to a particular IP address with `"/"` or you can direct it to things such as `"/docs"` and `"/admin"` as you choose. For this demonstration, set it to `"/"`:

    ```
    @app.route("/")
    ```

6. Finally, tell Flask what you want it to respond with. The default is HTML, so you can format your response string with HTML tags:

    ```
    def hello_world():
        return "<p> Hello World! </p>"
    ```

7. Run your Flask demonstration application using the following command:

    ```
    python - m flask --app FlaskTest run
    ```

    This produces the following output:

    ```
    * Serving Flask app 'FlaskTest'
    * Debug mode: off
    WARNING: This is a development server. Do not use it in a
    production deployment. Use a production WSGI server instead.
    * Running on http://127.0.0.1:5000
    Press CTRL+C to quit
    ```

8. In a browser window, go to `http://127.0.0.1:5000`. You'll see the following output in your browser:

Figure 13.4 – Web response from the "Hello World!" Flask application

In your Python window, you'll see messages from Flask responding to the request:

```
127.0.0.1 - - [27/Dec/2023 09:44:45] "GET / HTTP/1.1" 200 -
127.0.0.1 - - [27/Dec/2023 09:44:45] "GET /favicon.ico HTTP/1.1"
404 -
```

Flask will warn you that this is a development server and not for deployment in production. We'll address this warning later in this chapter by helping you put your Flask application into a Docker container with a **Web Server Gateway Interface (WSGI)** server.

> **Use production systems for deployment**
>
> When deploying your model into production, it's important to understand that Flask's development server is *not* intended for this purpose. The development server is designed for local use and lacks the critical security features and performance capabilities required for a production environment. It's not equipped to handle multiple simultaneous asynchronous requests, which your model will likely need to process in production.

To ensure your model runs efficiently, securely, and in a scalable and performant way, you should deploy it using Gunicorn and Nginx. Detailed examples of the use of Gunicorn and Nginx are beyond the scope of this book. That said, here's a brief overview of these two networking applications:

*   **Gunicorn** is a **WSGIHTTP server** designed for running Python web applications in production. It can handle **multiple concurrent requests** by managing multiple worker processes, making it highly scalable and capable of distributing the load efficiently across your CPU cores.

*   **Nginx** is a powerful **reverse proxy server** that sits in front of Gunicorn, handling client requests. It helps manage heavy traffic by serving static files and balancing requests efficiently across the Gunicorn workers. Nginx also adds an important layer of security and performance optimizations, such as handling SSL/TLS encryption, which Flask's development server doesn't support.

Now that you have an idea of how Flask works, you can use it to deploy the methods for each of the commands for your model. We'll do that now:

1. First, modify the file you created with the methods in them. Make sure you have the necessary packages, which include `Flask`, `jsonify` to send data back to the user, and `request` to get data from the URL. Start up Flask with `app=Flask(__name__)`:

```
import pandas as pd
from sklearn.metrics import mean_squared_error,
    r2_score
import pickle
import os
from flask import Flask, jsonify, request
app=Flask(__name__)
```

2. If the user wants to retrain the model, they'll call the API with `/retrain` and include JSON specifying the new dataset to use. This part of the API loads the model, accepts the JSON file from the URL, and calls the `.fit` method. It will output a message stating `"Model retrained successfully"` so that the user knows something happened:

```
@app.route("/retrain",methods = ["POST"])
def retrain_model(model="XBGpipeline.pkl"):
    #load the model from pickle file, retrain the model and save
to pickle
    with open(model, 'rb') as file:
        loaded_model_pickle = pickle.load(file)

    data = request.get_json()
    df = pd.DataFrame(data)

    # Split the data into features and target variable
    X_train = df.drop('target', axis=1)
    y_train = df['target']
    loaded_model_pickle.fit(X_train, y_train)

    base_name, extension = os.path.splitext(model)
    new_file_path = f"{base_name}_retrained{
        extension}"

    with open(new_file_path, 'wb') as file:
        pickle.dump(loaded_model_pickle, file)

    return jsonify({'message':
        'Model retrained successfully'})
```

3. If the user wants to make a prediction, they can do a similar thing by calling the API with `/predict` and including a JSON with the input values for the model to use in the prediction. The following code accepts the JSON and calls the `.fit` method, then outputs the prediction as a JSON using the Flask `jsonify` method:

```python
@app.route("/predict",methods = ["POST"])
def predict(model="XBGpipeline.pkl"):
    if request.method == "POST":
        X_test = request.get_json()

        #use the loaded model to predict a result
        with open(model, 'rb') as file:
            loaded_model_pickle = pickle.load(file)

        try:
            predictions_pickle =
                loaded_model_pickle.predict(X_test)

        except Exception as e:
            predictions_pickle = None
            print("An error occurred:", e)

        return jsonify({'predictions':
            predictions_pickle.tolist()})
```

4. To evaluate the model, the user can call `/eval` and pass a JSON with the prediction and actuals in pairs. This code accepts that JSON file and outputs evaluation metrics as a JSON file using a class called `EvalMetrics` and `jsonify`:

```python
@app.route("/eval",methods = ["POST"])
def evalmodel(model="XBGpipeline.pkl"):
    #requires json file with y_pred, y_actual
    data = request.get_json()
    df = pd.DataFrame(data)
    y_pred = df.drop('target', axis=1)
    y_actual = df['target']

    # Create a class to hold the results of model
    #evaluation mse, r_squared
    class EvalMetrics:
        def __init__(self):
            self.mse = 0
            self.r_squared = 0
```

```
EvalMetrics.mse = mean_squared_error(y_pred,
    y_actual)
EvalMetrics.r_squared = r2_score(y_pred, y_actual)
return jsonify({'Metrics': EvalMetrics.tolist()})
```

With that, you've used Flask to create an API for your model. With this API, a user can make a prediction, retrain the model, and evaluate the model's performance. Next, you'll learn how to use Docker to containerize your model and API, making it portable and easy to deploy on a cloud service.

Using Docker to put your model into a container

In this section, you'll learn how to containerize a model and API using Docker. Docker is a solution that separates an application from its infrastructure. With Docker, you can package your application in a portable container, which enables you to move it from one computer to another without the need for a lot of configuration work. It also separates your application from anything else running on the same hardware.

A quick overview of Docker

Docker is a client-server application. On the client system, you communicate with the Docker daemon to create and manage images and containers. The daemon runs on the server, which for development is typically the same system. With Docker, you can communicate with registries of images and use them as starting points for your applications. Docker also gives you tools to move your container from system to system. This is illustrated in *Figure 13.5*:

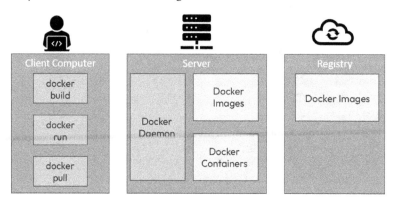

Figure 13.5 – An overview of Docker

Figure 13.5 shows that the various parts of the Docker system are in different locations. On the client computer, Docker gives commands to the Docker daemon running on the server. The docker build command creates a new image on the server, docker run executes an image in a container on the server, and docker pull gets images from a registry. Next, you'll create a Docker container that will hold your model, making your model portable to multiple cloud systems.

Building a container for your model

In this section, you'll learn how to set up the deployment software stack for your model. So far, you've been using Python and XGBoost for training and making predictions, with the model wrapped in a Flask API to allow external access. Now, to move this setup into a production environment, you'll need to add a few more components:

1. First, you'll need a WSGI server to run the Flask API. For this, you'll use Gunicorn, as discussed earlier.

2. Next, you'll need a web proxy server to manage client requests and route them inside the container efficiently. For this purpose, you'll use Nginx.

With these additions, your complete deployment stack will include Python, XGBoost, Flask, Gunicorn, and Nginx, all packaged into a Docker container for smooth production deployment.

Here's what the full stack will look like:

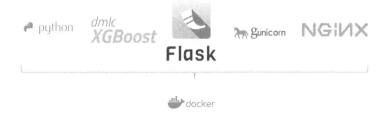

Figure 13.6 – Full software stack for your production model with an API

Figure 13.6, as mentioned previously, shows Python, XGBoost, Flask, Gunicorn, and Nginx all going inside a Docker container. That's exactly what happens, and it's what you'll build in the next section.

Deploying a model using containers

Before you can containerize your application, you need to ensure it runs properly in a production environment. Then, you can build a Dockerfile, a configuration file that reproduces that production environment as a container. You'll complete both the testing and containerization steps on your local system; then, in the next section, you'll learn how to deploy your container to the cloud.

Let's consider an example Dockerfile that includes the pieces you'd need to containerize the Flask API you created earlier in this chapter and the model from *Chapter 9*:

1. First, you'll need to inherit Python from a standardized image. You'll want to match the version to what you've been using – in this case, Python 3.9. The -slim option reduces the amount of disk space your container takes up:

```
FROM python:3.9-slim
```

2. Next, you'll want to establish a web socket so that it uses EXPOSE 8000, as well as do some setup so that you can see things happen as your container is run and keep things clean in the container. Here, ENV PYTHONUNBUFFERED 1 tells Docker not to buffer the output so that you see the output of the application in real time. Finally, ENV PYTHONDONTWRITEBYTECODE=1 keeps your container clean by telling Python not to make .pyc files:

```
EXPOSE 8000
ENV PYTHONUNBUFFERED 1
ENV PYTHONDONTWRITEBYTECODE=1
```

3. Next, you'll want to make sure that the applications inside the container are up to date by running apt-get update:

```
RUN apt-get update \
&& apt-get install -y --no-install-recommends python3\
&& apt-get clean
```

4. Now, you can make a directory in the container that will hold the API called /rest_api, set it as the working directory, and copy your Flask API to that directory:

```
RUN mkdir /rest_api
WORKDIR /rest_api
COPY . /rest_api/
```

5. Then, copy over the requirements.txt file so that all the Python dependencies for your model and API can be installed in the container:

```
COPY ./requirements.txt /rest_api/requirements.txt
```

6. Install the requirements while making sure wheel is updated first by using pip install:

```
RUN python3 -m pip --no-cache-dir install wheel
RUN python3 -m pip --no-cache-dir install -r requirements.txt
```

7. At this point, everything is ready, so you can run the wsgi file for your application using gunicorn, set up port 8000, and start three workers. When you use Docker to run your container, the command in CMD is executed, which will start gunicorn and run your API Flask application:

```
CMD ["gunicorn", "--bind", ":8000", "--workers", "3" "rest_api.
wsgi"]
```

With that, you've created a Dockerfile that you can use to deploy a containerized model to a registry and then retrieve the container on any cloud resource that runs Docker. Next, we'll discuss some aspects you should consider when deploying a container in the cloud.

Considerations when deploying a container in the cloud

The big cloud services (Google, Amazon, and Microsoft Azure) can run your Docker container and often have a free tier that you can use if you don't require storage or shared services. You'll need to copy your container Dockerfile to the cloud service and build the image in the cloud service. There are multiple cloud provider options you can use with a Docker container, each of which has its specific set of steps. For this reason, we won't cover it here. Feel free to try out the big cloud services or other containers as service providers. However, we will cover some important considerations when moving to production.

> **Considering risk and web security when deploying a model**
>
> Deploying a machine learning model into a production environment is a complex task that should be approached with caution. Beyond ensuring that your model performs as expected, any system that's exposed to the web is vulnerable to risks such as cyberattacks, hacks, and user errors. To minimize these risks, it's critical to follow DevOps best practices, which include performing rigorous input validation on all data received through your web API and thorough testing before deployment.

Ignoring best practices when deploying a model can expose sensitive data, compromise the security of your system, or lead to operational failures. While web security and operational concerns are beyond the scope of this book, we highly recommend diving into the following resources to gain a solid understanding of DevOps and MLOps for secure, scalable deployments:

- *The Phoenix Project: A Novel about IT, DevOps, and Helping Your Business Win*, by Gene Kim, et al.
- *DevOps for Web Development: Achieve Continuous Integration and Continuous Delivery of Your Web Applications with Ease*, by Mitesh Soni.
- *Engineering MLOps*, by Emmanuel Raj.

By leveraging the insights from these books, you can develop a deeper understanding of how to secure and optimize the deployment of machine learning models while minimizing risk.

In the next section, we'll discuss how you can service your model once it's been deployed using the REST API.

Servicing your model using the REST API

Once your model has been deployed and your container is running, you can use REST API commands to access it. These commands interface with the methods you created earlier in this chapter in the *Creating API calls to your model* section. In this section, you'll use REST API commands to interface with the time series model from *Chapter 9*. Let's begin by checking how the model is performing.

Monitoring model performance

To test how your model is performing, open a browser window and navigate to the URL that's serving the model – for example, say your model is hosted at `https://example.com`. Note that this is completely made up and this URL won't work. To check model performance, the appropriate REST command to use would be GET (`https://example.com/api/v1/metrics`). You can use the Python `requests` module to interact with the API programmatically. So, let's do that:

1. Begin by importing the `requests` module:

    ```
    import requests
    ```

2. Next, form the URL to GET the model performance metrics:

    ```
    url = https://example.com/api/v1/metrics
    ```

3. Now, set up headers so that you can pass your API key:

    ```
    headers = { "api_key": "YOUR_API_KEY" }
    ```

4. Then, create a variable to store the response:

    ```
    response = requests.request("GET",url,
        headers=headers)
    ```

5. Finally, print the response from the API:

    ```
    print(response.text)
    ```

Next, you'll learn how to check the model metrics using the API.

Determining when to retrain

Knowing when to retrain your XGBoost model is key to keeping its predictions accurate over time. By tracking performance metrics such as accuracy, precision, recall, and RMSE, you can spot signs of decline. Changes in data patterns, feature distributions, or drops in prediction quality often signal that it's time for retraining. Regularly comparing recent data to your original training data and monitoring these metrics helps you set thresholds that indicate when the model is no longer performing well. This ensures your model stays reliable and accurate in production.

Retraining your model

Retraining your model means updating it with new data to reflect changing trends and patterns. This process begins by gathering fresh, labeled data that represents the current environment in which your model will be working. From there, you'll preprocess this new data, merge it with existing datasets, and adjust hyperparameters to maintain or even improve performance. Once the model has been retrained, it's crucial to validate it against your benchmarks before redeploying it to production, ensuring it continues to provide accurate, reliable predictions.

In this section, we discussed how you can use the REST API to interact with your model once it's been deployed into production. You saw how to perform common maintenance tasks such as monitoring performance metrics. Now, let's recap what we learned in this chapter.

Summary

In this chapter, you learned how to deploy your XGBoost model into production. You built an API to allow users to access the model through a web interface and make predictions, known as inference. You also created functions to manage your model by adding new data and retraining it when needed. Additionally, you explored how to monitor model performance using the metrics discussed in *Chapter 11*, and you learned how to determine when retraining is necessary.

You started this chapter by exploring multithreaded training options and wrapped up with cloud-based deployment using containers.

At this point, you've gained a solid understanding of how to use XGBoost for various types of data. You've had hands-on experience with the XGBoost Python API through practical use cases in classification, regression, and time series analysis. You've also practiced testing, evaluating, and deploying your models into production.

We hope this book has provided you with a strong understanding of the XGBoost algorithm, helped you successfully install and use the XGBoost API, and guided you through the process of preparing datasets, training models, making predictions, and evaluating and deploying them using **Python** and **scikit-learn**. We wish you continued success in your journey with machine learning and XGBoost!

Index

packtpub.com

Subscribe to our online digital library for full access to over 7,000 books and videos, as well as industry leading tools to help you plan your personal development and advance your career. For more information, please visit our website.

Why subscribe?

- Spend less time learning and more time coding with practical eBooks and Videos from over 4,000 industry professionals

- Improve your learning with Skill Plans built especially for you

- Get a free eBook or video every month

- Fully searchable for easy access to vital information

- Copy and paste, print, and bookmark content

Did you know that Packt offers eBook versions of every book published, with PDF and ePub files available? You can upgrade to the eBook version at packtpub.com and as a print book customer, you are entitled to a discount on the eBook copy. Get in touch with us at customercare@packtpub.com for more details.

At www.packtpub.com, you can also read a collection of free technical articles, sign up for a range of free newsletters, and receive exclusive discounts and offers on Packt books and eBooks.

Other Books You May Enjoy

If you enjoyed this book, you may be interested in these other books by Packt:

Generative AI Foundations in Python

Carlos Rodriguez

ISBN: 978-1-83546-082-5

- Discover the fundamentals of GenAI and its foundations in NLP

- Dissect foundational generative architectures including GANs, transformers, and diffusion models

- Find out how to fine-tune LLMs for specific NLP tasks

- Understand transfer learning and fine-tuning to facilitate domain adaptation, including fields such as finance

- Explore prompt engineering, including in-context learning, templatization, and rationalization through chain-of-thought and RAG

- Implement responsible practices with generative LLMs to minimize bias, toxicity, and other harmful outputs

MATLAB for Machine Learning - Second Edition

Giuseppe Ciaburro

ISBN: 978-1-83508-769-5

- Discover different ways to transform data into valuable insights
- Explore the different types of regression techniques
- Grasp the basics of classification through Naive Bayes and decision trees
- Use clustering to group data based on similarity measures
- Perform data fitting, pattern recognition, and cluster analysis
- Implement feature selection and extraction for dimensionality reduction
- Harness MATLAB tools for deep learning exploration

Packt is searching for authors like you

If you're interested in becoming an author for Packt, please visit `authors.packtpub.com` and apply today. We have worked with thousands of developers and tech professionals, just like you, to help them share their insight with the global tech community. You can make a general application, apply for a specific hot topic that we are recruiting an author for, or submit your own idea.

Share Your Thoughts

Now you've finished *XGBoost for Regression Predictive Modeling and Time Series Analysis*, we'd love to hear your thoughts! Scan the QR code below to go straight to the Amazon review page for this book and share your feedback or leave a review on the site that you purchased it from.

https://packt.link/r/1-805-12305-X

Your review is important to us and the tech community and will help us make sure we're delivering excellent quality content.

Download a free PDF copy of this book

Thanks for purchasing this book!

Do you like to read on the go but are unable to carry your print books everywhere?

Is your eBook purchase not compatible with the device of your choice?

Don't worry, now with every Packt book you get a DRM-free PDF version of that book at no cost.

Read anywhere, any place, on any device. Search, copy, and paste code from your favorite technical books directly into your application.

The perks don't stop there, you can get exclusive access to discounts, newsletters, and great free content in your inbox daily

Follow these simple steps to get the benefits:

1. Scan the QR code or visit the link below

https://packt.link/free-ebook/9781805123057

2. Submit your proof of purchase
3. That's it! We'll send your free PDF and other benefits to your email directly

www.ingramcontent.com/pod-product-compliance
Lightning Source LLC
LaVergne TN
LVHW081518050326
832903LV00025B/1529